6th Biennial Conference of the Interdisciplinary Society for Quantitative Research in Music and Medicine (ISQRMM 2021)

Published in the Journal of the Interdisciplinary Society for Quantitative Research in Music and Medicine

Volume Number 6 (2021)

Online
19 – 20 June 2021

ISBN: 978-1-7138-6647-3

Printed from e-media with permission by:

Curran Associates, Inc.
57 Morehouse Lane
Red Hook, NY 12571

ISBN: 978-1-7138-6647-3

Printed with permission by Curran Associates, Inc. (2023)

ISQRMM
C/O David O. Akombo, Ph.D.
Dean, Faculty of Culture, Creative and Performing Arts
P.O. Box 64
Bridgetown BB11000
Barbados, West Indies

http://isqrmm.org

The *Journal of the Interdisciplinary Society for Research in Music and Medicine* is a peer-reviewed journal presenting scholarly research into music's connection and application in medicine. Articles are submitted from scholars who have presented at the society's biannual conference, as well as from the field. Inquiries regarding submission should be directed to contact@isqrmm.com

Additional copies of this publication are available from:

Curran Associates, Inc.
57 Morehouse Lane
Red Hook, NY 12571 USA
Phone: 845-758-0400
Fax: 845-758-2634
Email: curran@proceedings.com
Web: www.proceedings.com

Table of Contents

President's Message

The study of the beneficial effects of music in medicine is a discipline that has spurred an interest among scholars since the dawn of the 18th century. Scholars from many spheres of human development have been intrigued but especially those from the humanities and biomedical sciences. Researchers in the 19th and 20th centuries continued to embrace these studies with scholarly and academic vigor paving the way for those in the 21st century to explore it even deeper. The mission of the *Journal of the Interdisciplinary Society for Quantitative Research in Music and Medicine* is to propel this scholarship with vigor into the 21st century and provide a global readership of peer reviewed articles that explore the unanswered questions regarding the efficacies of all forms of music on the human body, mind, and soul. The *Journal of the Interdisciplinary Society for Quantitative Research in Music and Medicine* welcomes qualitative research as well as quantitative research methodologies and encourages interdisciplinary formats. The Journal encourages the authors to utilize modes of research that are clearly defined, explained, and relevant. The Journal also welcomes Master's or Doctoral theses in any related fields.

The topics of historical significance impinging upon qualitative research on music and healing, especially as revealed in its manifold and remarkable manifestations—throughout Ancient Greece in the 18th century inspiring scholarship around ancient civilizations—has never been candidly and fairly examined in the lenses of the contemporary scientific models. To this end, there is a dearth of research spanning the global cultures that use music as non-phrenological medicine are worthy of this examination. Presently, the only sound sources of information on the culturally diverse aspects of music as medicine are anecdotes, which provide eyewitness accounts of individual music healers. In addition to the quantitative studies in the field, the anecdotal repository of evidence, as recorded in treatises of healing rituals, add credence to the historical implication of music as a healing agent.

The *Journal of the Interdisciplinary Society for Quantitative Research in Music and Medicine* contains cutting-edge research that balances its content with readership that is interdisciplinary in scope, while using eclectic models to present the research findings. These make the journal both scholarly and informative for all readers regardless of specialization. We are proud of ISQRMM community, enthusiasts, and erudite scholars that make up our global research clientele.

Musically Yours,

Professor David O. Akombo, PhD, M.A., M.Mus., B.Ed.

Co-Founder and President

Co-Founders

David O. Akombo, Ph.D.

Joseph Ruszkowski, Ph.D.

Estate of Arthur Harvey, DMA (Professor of Music and Psychology, University of Hawai)

Estate of Jorge Camara, M.D. (Professor of Ophthalmology, University of Hawaii School of Medicine)

Board Members

David O. Akombo, Ph.D. (Professor of Music Education and Interdisciplinary Musicology)

Steve Jackowicz, Ph.D. (Professor of Chinese Medicine)

Kent Nelson, DMA, (Independent Scholar)

Elise Zotos, (Practitioner of music for health)

Adina Dabija, LAc, MSTOM, Dipl OM

Editor's Forward

Dear readers, researchers, and members of the ISQRMM Board, I want to welcome you to the 6th volume of the Journal of the Interdisciplinary Society for Quantitative Research in Music and Medicine (ISQRMM). The past two years have been extraordinarily trying times due to the Covid-19 pandemic and the necessity of holding our biannual conference in 2021 virtually for the first time since the initial conference held in Ogden, Utah, USA in 2011. The mission of ISQRMM and the consequential studies done by researchers in the field of music and medicine is needed now more than ever. Added to the urgency of the pandemic was and is the opioid crises affecting many individuals, families, and communities. Research in complementary and alternative methods of treatment without using pharmaceutical drugs may be paramount. The use of music for healing may hold key health benefits for many individuals. Research is warranted to tease out these benefits and the articles in this journal are such an approach.

At times it is necessary to look backwards as we look forward to progress with our understanding of healing. The articles by Dr. Adina Dabija and Antonios Chrysovergis give us an historical view of healing in the ancient world that may hold benefits for today. With an eye to the future, Dr. June Weider shared her research on the use of music and vibration on the spine. Dr. David Akombo is proposing research to examine the possible benefits of music in the recovery of Covid-19 patients. The use of music and community has been researched by Jonathan Segilman and Rona Geffen. In the case of Segilman's article, it was the use of music to help a school and community grieve for a teacher that had passed away; Geffen explored the self-healing that may occur for individuals using vocalization at parties and rave events. Dr. Sheerin Hosseini researched the frustrations, benefits, and positive qualities of individuals learning music with a learning disorder (dyscalculia). It is the strength of ISQRMM the organization and conference that brings together such a diverse area of interests under one banner: music, the brain, health, and healing.

Please read, ponder, and enjoy reading the 2021 ISQRMM Journal as we explore the various subjects related to research in music and the mind and body. Music affects individuals in different ways. If, through research, we can find health benefits outside the usual accepted practices of casual music listening and performing, or through the careful study of the nuances of everyday musicing, this is a positive in the complex world of today. The members of the ISQRMM Board invite you to attend our future conferences and submit articles for publication. The organization is wide in scope with many eclectic interests. Welcome!

Kent P. Nelson, DMA
Editor and Board Member

Co-editors:
David O. Akombo, PhD
Adina Dabija, LAc, MSTOM, Dipl OM
Elise Zotos

The *Journal of the Interdisciplinary Society for Research in Music and Medicine* is a peer-reviewed journal presenting scholarly research into music's connection and application in medicine. Articles are submitted from scholars who have presented at the society's biannual conference, as well as from the field. Inquiries regarding submission should be directed to contact@isqrmm.com

Music in Ancient China: Correlations Between 五 音 (*wu yin*) – The Five Nodal Sounds, 五 行 (*wu xing*) - The Five Phases, 十天干 (*shi tian gan*) - The Ten Celestial Stems and Human Health in 论五音建运 (Lun Wu Yin Jian Yun) - *Discourse on the Five Notes Health (Cosmic) Movements* by Xu Chunfu (1556)

Adina Dabija

Abstract

The correlation between the movement of the astral bodies and the phenomenal world is a fundamental concept in Asian culture. Like in many other ancient cultures, such as Hellenism, Hinduism, or the Hermetic tradition, holographic correspondences were traced between the whole and its parts. In an effort to identify the underlying patterns of the cosmos and the ideal way of governing people, Asian culture generated many records on how the movements of the celestial bodies influenced the life on Earth and how to navigate these influences. In this paper I am investigating correlations between the astrological dynamics generated by the Ten Celestial Stems (十天干 - shi tian gan), the Five Earthly Phases (Wood, Fire, Earth, Metal, and Water) (五行 - wu xing), the five musical notes of the pentatonic scale (五 音 - wu yin) and human health based on my translation of the *Discourse on the Five Notes Health (Cosmic) Movements* (论五音建运 - Lun Wu Yin Jian Yun) by Xu Chunfu, originally published in 1556.

Keywords: Asian Medicine, Ten Celestial Stems, Five Phases, Five Nodal Sounds, Harmony

Historical Background

The Ten Celestial Stems (十天干 - *shi tian gan*) – is an aspect of the Chinese astrology older than the concept of Five Phases (Wood, Fire, Earth, Metal, and Water) (五 行 - *wu xing*) - as it appears in early oracular writings in the Shang dynasty (1,700-1,046 B.C.E.). The concept was originally developed by the 商朝 (*shang*) - "high" or "superior" people who identified The Ten Stems with their mythical ancestors, from whom they inherited the 天命 (*Tianming*) - "the mandate of heaven" to gain authority and rule in the bronze age China. This concept evolved into the astrological ten great suns during the Zhou Dynasty, with Shang Di - "the high lord" as the "pivot of the Heavens," believed by many translators to be the Polar Star, the brightest star of Ursa Minor constellation. Chapter 67 from *Huang Di Nei Jing, Su Wen* (Simple Questions), attributed to Huang Di, the Yellow Emperor (2,696 – 2,598 B.C.E.) says,

> In order to understand the origin of the phase energetic system, one must possess knowledge of astronomy. The planets within our galaxy exert the most influence on phenomena in our world. There are further twenty-eight constellations that are observable with the naked eye in the heavens that are also significant to human life. The constellations also span the entire sky, adding up to 360 degrees. With the North Star as the center of the circle in the sky, one can identify the four directions and hence each of the corresponding groups of constellations.

This quote reveals that the concept of 十天干 (shi tian gan) - the ten celestial stems might be older than the concept of 五 行 (wu xing) - five phases (Wood, Fire, Earth, Metal, and Water), or at least that, in the times of the Yellow Emperor, the influences of the constellations were seen as being at the origin of the climatic changes on earth, generating the five seasons (spring, summer, late summer, fall, and winter).

Chapter 71 from Ling Shu (靈樞經 - *The Spiritual Pivot*) mentions the relationship between the 五音 (wu yin) – the five tones, 五 行 (wu xing) - the five phases and 十天干 (shi tian gan) - the ten celestial stems. The text was compiled in twelve scrolls during the Warring

States (480-222 B.C.E.) from oral tradition and mentions “mutual resonances between man and heaven and earth” such as, “Heaven has five tones. Man has five viscera,” “Heaven had the ten days of the celestial stems. Man’s hands have ten fingers,” and “Heaven has yin and yang. Man has male and female.” Hence, one can draw the conclusion that the musical esthetics in Ancient China mirrored the energetic aspects of the astrological influences of “heaven” and was charged with symbolic meaning.

Such correspondences were later refined in the text translated and discussed below, authored by Xu Chunfu, a doctor who served in the imperial medical bureau during the Ming dynasty. One of the interesting things we know about Xu Chunfu is that he established one of the first medical associations for doctors, the Medical Society for Harboring Kindness in 1568 in Beijing, with forty six members that signed a pact agreeing to abide by ethical and technical guidelines. This text is part of a collection of articles, 古今医统大全 - *Gu Jin Yi Liu Da Quan (The Great Compendium of Ancient and Modern Medical Traditions)* published by Xu Chunfu in 1556, including medical knowledge from over three hundred published and unpublished Chinese medical classics. As an imperial doctor, Xu Chunfu had access to a wide body of medical literature. The collection was probably meant as an instrumental resource or textbook for the members of the medical society founded by Xu Chunfu.

Historically, there are two different systems of correlations between the Ten Celestial Stems and the Five Phases in Asian culture. The customary one results from the division of the ten stems to the four cardinal directions, with the earth in the center. This is not the system used in the text I translated, although I will present it here for your reference - the table below depicts the assignment of the 5 dynamics, yin and yang to the heavenly stems with their corresponding organs:

Stem	Pinyin	Pictogram	Direction	ZangFu
S.I	Jia	甲	East	GB
S.II	Yi	乙		LV
S.III	Bing	丙	South	SI
S.IV	Ding	丁		HT
S.V	Wu	戊	Center	ST
S.VI	Ji	己		SP
S.VII	Geng	庚	West	LI
S.VIII	Xin	辛		LU
S.IX	Ren	壬	Nord	BL
S.X	Gui	癸		KI

The correlation system used in Xu Chunfu's text was first described in chapter 67 of Huang Di Nei Jing, *Su Wen* (Simple Questions). It was developed from observing meteorological phenomena - wind, heat, damp, summerheat, dryness, and cold – in correlation with the "movement of the stars" and their impact on Earth. The transformational phases (elements) follow each other in the generation *(sheng)* cycle. Two stems are paired with each other, in accordance to the principle of polarity *yin* 陰/ *yang* 陽 - day and night, growth and decay, male and female – as referred in Chapter 42 from 道德經 Dao De Jing (around 200 B.C.E.), dedicated to the divisions of the Dao (the mysterious principle). This system is applied to climatology and astrological calculations.

Elements Motion (Yun Xing)	Heavenly Stems	
Earth	I	VI
Metal	II	VII
Water	III	VIII
Wood	IV	IX
Fire	V	X

论五音建运 Lun Wu Yin Jian Yun "Discourse on the Five Notes Health (Cosmic) Movements"

土曰宫，其位甲己之岁。
Tǔ yue gong, qi wei jia ji zhi sui.
The **Earth/Soil** called *Gong* (palace), resonates with the quality of the *Jia* (1st) and the *Ji* (6th) Heavenly Stems.

宫，土也，中和之道，无往而不理。
Gong, tǔ yě, zhōng hé zhī dào, wú wǎng ér bù lǐ
Gong/soil is the way of central harmony if one doesn't enter (this space), than there is no order or understanding.

又总堂室奥阶而谓之宫，所闱不一，
You zong tang shì ào jiē ér wèi zhī gong, suǒ wéi bù yì.
Also, the complex of halls, rooms, dark corners, and stairs is also called gong/palace and does not have only one entrance.

盖土亦以通贯于金木水火，
Gài tǔ ban yǐ tōng guàn yú jīn mù shuǐ huǒ,
The envelopment of the Earth also connects and interpenetrates Metal, Wood, Water, and Fire,

主于四季，荣于四脏，皆总之之意。
zhǔ yú sì jì, róng yú sì zāng, jiē zǒng zhī zhī yì.
hosts the four seasons, enables the four organs to flourish, this is the overall meaning.

故五运从十干起甲，为土也。
gù wǔ yùn cóng shí gàn qǐ jiǎ, wéi tǔ yě 。
Therefore, the five movements follow the Ten Celestial Stems beginning with Jia (HS6, Wood Yang), which corresponds to Earth/soil.

土生金，故乙次之。
tǔ shēng jīn, gù yǐ cì zhī。
Earth generates **Metal,** therefore *Yi* heavenly stem (HS6, Earth yin) is next.

金生水，故木次之。
jīn shēng shuǐ, gù mù cì zhī
Metal generates Water; therefore, Wood is next.

如此五行相生而转，甲为阳，乙为阴，亦相间而数，如环无端。
rú cǐ wǔ xíng xiāng shēng ér zhuǎn, jiǎ wéi yang, yǐ wéi yīn, yì xiāng jiān ér shù, rú huán wú duān
This is how the Five Phases mutually generate and turn. Jia stem acts as yang, Yi serves as (becomes) yin. They are next to each other and thus counted: in this way they cycle without end.

在阳年曰太，在阴年曰少；
zài yáng nián yuē tài, zài yīn nián yuē shào
In the yang year it is called *tai* (great), in a yin year is called *shao* (lesser)

阳年大过为司天所抑，阴年不及为年支所合，皆曰平气。
yáng nián dà guò wéi sī tiān suǒ yì, yīn nián bù jí wéi nián zhī suǒ hé, jiē yuē píng qì.
The excess of the yang year is controlled by the Driver of Heaven; the deficiencies of the yin year are united by the earthly branch of the year, and these are referred by normal/ calm qi.

金曰商，其位乙庚之岁。
jīn yuē shāng , qí wèi yǐ gēng zhī suì 。
(The note for) Metal is *Shang*. Its position is in the Yi (HS2, Wood Yin) and *Geng* (HS7, Wood Yang).

商，强也，谓金性之坚强也。
Shāng, qiáng yě, wèi jīn xìng zhī jiānqiáng yě.
Shang is strong as it has the hard strengths of Gold/Metal.

水曰羽，其位丙辛之岁。
shuǐ yuē yǔ, qí wèi bǐng xīn zhī suì
The note of **Water** is *Yu*. Its position is in the *Bing* (HS 3, Fire Yang) and *Xin* (HS 8, Metal Yin) years.

羽，舒也，阳气将复，万物孳育而舒生也。
Yǔ, shū yě, yáng qì jiāng fù, wàn wù zī yù ér shū shēng ye。
Yu means to unfold (stretch out), Yang qi is about to return and ten thousand things breed, are raised and generation unfolds.

木曰角，其位丁壬之岁。角，触也，象诸阳气触动而生也。
Mù yuē jiǎo, qí wèi dīng rén zhī suì. Jiǎo, chù yě, xiàng zhū yáng qì chùdòng ér shēng yě.
The note for **Wood** is *Jue*. Its position is in the *Ding* (HS 4, Fire Yin) and *Ren* (HS9, Water Yang) years. Jue means to butt or touch; it is like all Yang Qi - to touch, move, and generate.

火曰征，其位戊癸之岁。征，止也，言物盛则止也。
Huǒ yuē zhēng, qí wèi wù guǐ zhī suì. Zhēng, zhǐ yě, yán wù shèng zé zhǐ yě.
The note of **Fire** is *Zhi*. Its position is *Wu* (HS5, Earth Yang) and *Gui* (HS 10, Water Yin).
Zhi means to stop. When something is full/ complete, it halts.

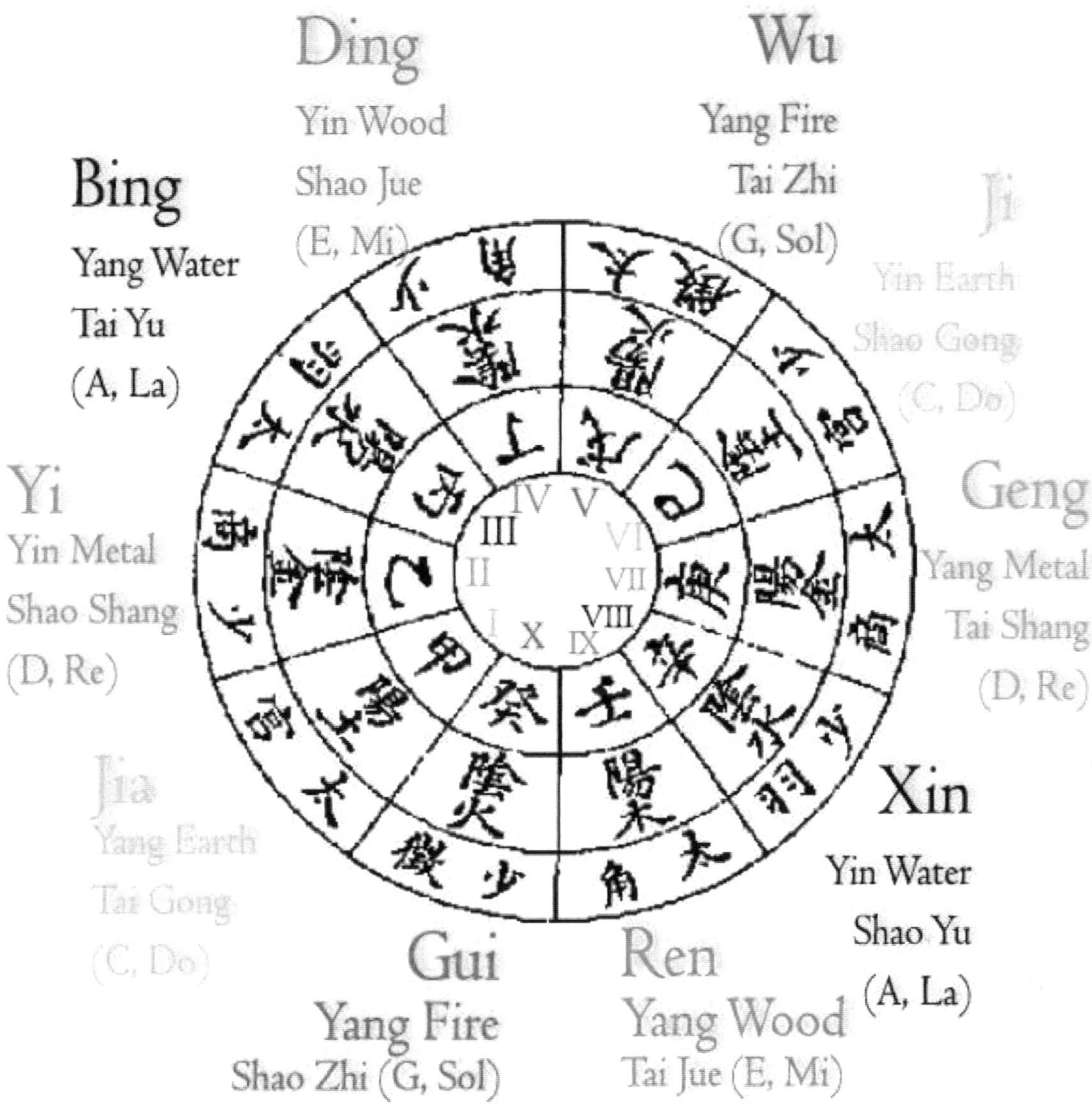
Ding
Yin Wood
Shao Jue
(E, Mi)
Wu
Yang Fire
Tai Zhi
(G, Sol)
Bing
Yang Water
Tai Yu
(A, La)
Ji
Yin Earth
Shao Gong
(C, Do)
Yi
Yin Metal
Shao Shang
(D, Re)
Geng
Yang Metal
Tai Shang
(D, Re)
Jia
Yang Earth
Tai Gong
(C, Do)
Xin
Yin Water
Shao Yu
(A, La)
Gui
Yang Fire
Shao Zhi (G, Sol)
Ren
Yang Wood
Tai Jue (E, Mi)
I
II
III
IV
V
VI
VII
VIII
IX
X

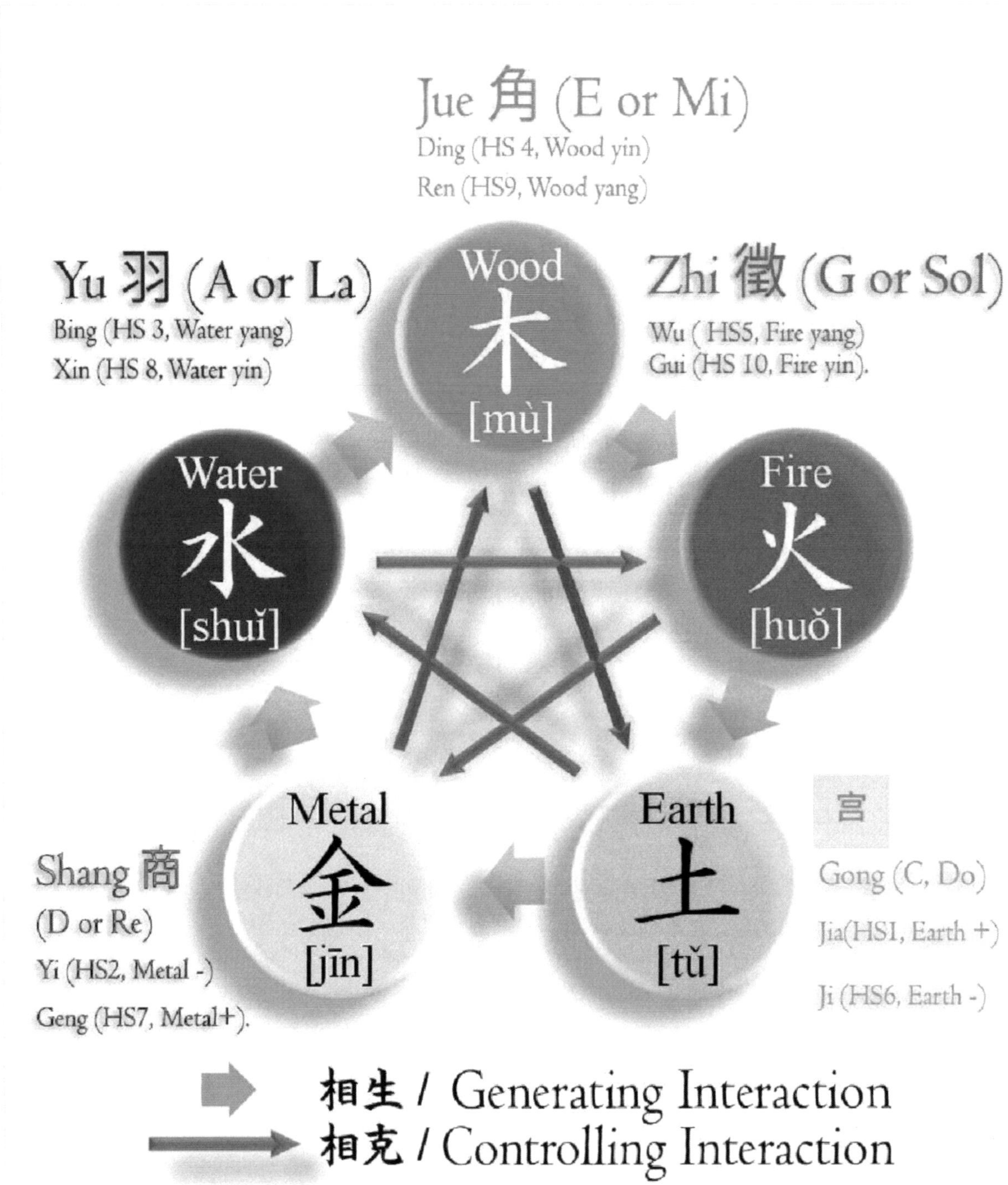
Jue 角 (E or Mi)
Ding (HS 4, Wood yin)
Ren (HS9, Wood yang)
Yu 羽 (A or La)
Bing (HS 3, Water yang)
Xin (HS 8, Water yin)
Wood
木
[mù]
Zhi 徵 (G or Sol)
Wu (HS5, Fire yang)
Gui (HS 10, Fire yin).
Water
水
[shuǐ]
Fire
火
[huǒ]
Metal
金
[jīn]
Earth
土
[tǔ]
宫
Gong (C, Do)
Jia(HS1, Earth +)
Ji (HS6, Earth -)
Shang 商
(D or Re)
Yi (HS2, Metal -)
Geng (HS7, Metal+).
相生 / Generating Interaction
相克 / Controlling Interaction

Discussion and Clinical Aspects

Harmonizing with 天 (Tian) - heavens was considered a necessary step in achieving the most efficient functioning of the human being in both of the main religious systems of China - Daoism and Confucianism. The practices of tuning in and aligning with Heaven through practicing virtue were encompassed in the Daoist concept of 養生 (yang sheng) - nourishing life, an important tool in preserving one's mind and body into old age, essentially buying time to complete one's spiritual evolution. During the Neo-Confucian Ming dynasty, when this text was written, proper or upright 正 (*zheng*) music was encompassed in the Confucian concept of 宜 (*yí*) – appropriateness. Thus, music was considered a tool of man's alignment with the forces of Heaven and Earth.

Each of the five notes - Gong (C or Do) – corresponding to Earth and resonating with the Jia/ Ji stems, Shang (D or Re) – corresponding to Metal and resonating with Yi/ Geng stems, Jue (E or Mi) – corresponding to Wood and resonating with Ding/ Ren stems, Zhi (G or Sol) – corresponding to Fire and resonating with Wu/Gui stems and Yu (A or La) – corresponding to Water and resonating with Bing/Xin stems – has specific energetics and intensities, according to the yin or yang environmental dynamic. These dynamics are reflected within the human body, activating certain climatic conditions that can give rise to disease. The five notes could be applied to correct such climatic features. For instance, Gong is the sound of Yang Earth and Yin Earth. The Yang Earth corresponds to Tai Gong (greater gong) and the Yin Earth to Shao Gong (lesser gong). Tai Gong generates Yin Metal, so Yin Metal type diseases could be addressed musically by playing a strong Gong to tonify the mother element of Metal, which is Earth. Thus, the fives sounds could be systemically applied to analyze the temporal and spatial variations of human illnesses and promote health.

One of the important factors that can change the internal climate in the body, according to Asian Medicine and also to modern psychosomatic medicine, are emotions. Negative emotions can generate diseases that can weaken the 氣 (qi) - vitality of their corresponding organs.

Anger corresponds to the phase of Wood and weakens the liver. Songs in the key of Jue (E or mi) are vigorous, melodious, cheerful, and bright – the music of the spring. Anger drives qi upwards, leading to rebellions of liver qi. Music in the key of Jue has the function of uplifting and inspiring and if there is qi stagnation - soothing liver qi. Maybe this is why Xu Chunfu's text mentions in respect to Jue: "it is like all Yang qi - to touch, move and generate."

Joy/happiness corresponds to the Fire element and is related to the proper functioning of the heart. The songs written in this key are warm, cheerful, relaxing, and contagiously joyful. Songs in the key of Zhi (G or Sol) are the music of the summer. By contrast, frenzy scatters the Shen (mind), leading to restlessness, palpitations, insomnia and insanity. While joy within normal range facilitates the circulation of blood and uplifts the spirit. Thus, Zhi songs invigorate heart; Yang, expel cold and regulate Qi and blood. Xu Chunfu's text refers to the fullness and the completion of this note.

Thoughtfulness corresponds to the Earth element. Songs in the key of Gong (C or Do) are characterized by being calm, solemn, mellow, and heavy. According to the Five Phases theory, this is the music of the late summer. When someone overthinks or ruminates, qi is bounded and digestive problems may appear. The sound of Gong strengthens the spleen and stomach and improves the ascending and descending of the qi dynamic of the middle jiao. Digestion is considered to be a core aspect of one's health and this might be why in Xu Chunfu's music theory "Gong/ soil is the way of central harmony if one doesn't enter (this space), than there is no order or understanding."

Sorrow corresponds to Metal and is related to the lungs. The songs in the key of Shang (D or re) are characterized by being resounding, sorrowful, solemn, sonorous and forceful. This is the music of autumn. However, sorrow consumes qi. The music of Shang strengthens the astringing and purifying effects of the lungs and regulates water metabolism. Xu Chunfu's text says: "Shang is strong as it has the hard strengths of Gold/ Metal."

Fear corresponds to the Water element and affects the kidneys. This category is characterized by being pure, plaintive, desolate, and gentle. It could be called the music of winter. The music of Yu (A or la) helps the kidneys to consolidate qi and essence, improves memory, and prevents aging.

Conclusion

Xu Chunfu's text offers a basic overview of the complex relationships between the Ten Celestial Stems, the Five Phases and the five notes as understood at the time of the Ming dynasty. We can view this text as an old and sophisticated form of music therapy and chronobiological medicine, in which the seasonal and yearly rhythms are responsible for regulating rhythms within the human biological clock that are mostly unknown or ignored in modern times. A better connection of man to the environment could be achieved by a deeper investigation of these correlations.

References

Allan, S. (2007). On the identity of Shang Di 上帝 and the origin of the concept of a celestial mandate (Tian Ming 天命). *Early China, 31*, 1-46. Retrieved from http://www.jstor.org/stable/23354211

Ben H, et al. (2010). *Clinical application and research on ziwu-liuzhu (midnight-noon ebb-flow) acupuncture therapy*; 35(3), 229-31. Chinese. PMID: 20848902.

Hinrichs, TJ., & Barnes, L. L. (Eds.). (2013). *Chinese medicine and healing: An illustrated history.* Belknap Press: An Imprint of Harvard.

Jarrett, L. S. (1998). *Nourishing destiny, the inner tradition of Chinese medicine*. Massachusetts, USA: Spirit Path Press.

Major, J. S. (1994). *Celestial cycles and mathematical harmonics in the "Huainanzi."* Extrême-*Orient Extrême-Occident, 16,* 121-134. Retrieved from http://www.jstor.org/stable/42635688

Zhang, H., & Lai, H. (2017) *Five Phases Music Therapy (FPMT) in Chinese Medicine: Fundamentals and Application*. Open Access Library Journal, 4, 1-11. doi: 10.4236/oalib.1104190.

An Original Harmonic and Scientific Approach to the Pythagorean Harmony of the Spheres and its Practical Application in Music Therapy

Antonios Chrysovergis

Abstract

The purpose of this study is to preserve, revive, and present the ancient Pythagorean philosophical and theological idea known today as Harmony of the Spheres, originally made known to us by Plato in the myth of Er, in Book X of the Republic, and to reintroduce it as a working system. In favor to achieving this, information is drawn exclusively from the ancient Greek tradition, with particular focus on the Pythagorean and Platonist philosophers.

For the application of this original system, the various musical scales implied or even proposed directly by the aforementioned philosophers are employed as a correspondence to the planetary sounds. Moreover, one original composition on the ancient Greek lyre and various music examples are included to underline the system. Furthermore, the moral theory of Plato, the idea that ancient Greek scales invoke certain types of ethos, the Pythagorean Tetractys, live astronomy, as well as ancient astrological models, are utilized to highlight the structure.

Finally, through extensive personal research using the system, further optimization techniques are proposed for this system to be adjusted beneficially to modern music and the scientific field of music therapy.

Keywords: Harmony, Terpander, Orpheus, Pythagoras, Plato

Introduction

Today, the association of the seven classical planets with musical pitches originally made known to us by Plato is primarily attributed to the Greek philosopher Pythagoras. Historically speaking, Pythagoras was the first philosopher who approached *mousikē* (music) scientifically, as we understand the term today. For the ancient Greeks, *mousikē* was "any art over which the Muses presided" (Liddel et al., 2011, p. 1148) and was not limited to the art of organized sound. Conversely, it would be more accurate to view Pythagoras as a link in a long chain within the framework of the Ancient Greek tradition, rather than a stand-alone innovator of advanced mathematical, musical, and cosmological models. Furthermore, to better understand the Pythagorean Harmony of the Spheres as part of the continuation, it would be helpful to further examine it in the context of the Greek Mystery traditions, specifically the Orphic tradition that Pythagoras was so inexplicably connected with, along with the tradition of Plato.

A New Light on the Harmony of the Spheres
On The Divine Origin of *Harmonia* ("Harmony")

The Harmony of the Spheres, as it appears in the ancient Greek tradition, is an idea that is intimately connected with the notions of the divine origin of *musike* (music, poetry, and dance) and more specifically with *harmonia* (harmony), also understood as the orderly arrangement of nature and the Cosmos. For the people of ancient Greece, music was a gift of the Gods to mankind. On the Greek island of Samothrace located in the northwestern Aegean Sea, Goddess Harmonia, the personification of idealized harmony and the daughter of God Zeus and Electra, was celebrated as one of the Great Gods in the mystery rites that took place on the island by the cult of the Great Gods (Siculus, 1939, p. 231). In the Orphic hymn "To Pan," we witness the God acting in the universe as a "weaver of playful song" who creates a "song of cosmic harmony" (Athanassakis & Wolkow, 2013, p. 13). Similarly in the Orphic Hymn "To Apollon," we are informed that the God of light is in possession of a "golden lyre" which He uses to infuse "harmony into the lot of men" and to regulate the rhythm of the seasons (Athanassakis & Wolkow, 2013, p. 30-31). Homer begins both the Iliad and the Odyssey with invocations to the "Goddess" (Homer, 1924, p. 13) and the "Muse" ("Homer, 1919, p. 13) respectively, with both being a probable reference to the muse Calliope, protector of epic poetry in ancient Greece. Furthermore, some of the most iconic ancient Greek instruments are said to have been invented by the Gods. In the Homeric hymn to God Hermes, the pastoral God invents the seven-string lyre (Homer, 2003, p. 117, line 39-55) and, in his effort to reconcile with his brother Apollo after a dispute over a herd of stolen oxen, he gives the lyre to him as a gift (Homer, 2003, p. 151-153, line 490-496). In other myths related to Orpheus' parentage, it is God Apollo himself who passes down his lyre to Orpheus, a hero, a founder of mystery rites, and a musician of unparalleled skill. Literary references such as these are a living testament of the creative capacity of our ancient ancestors to communicate the notion of *mousikē*'s divine origin.

On the Meaning of the Word *Harmonia* and Early References of the "Harmony of the Spheres"

The Greek word *ἁρμονία* ("harmonia," harmony) appears for the first time in Homer's Odyssey, *καὶ ἁρμονίῃσιν ἄρηρεν,* "and fitted them to one another" (Homer, 1919, p. 201) and in the context of his text is understood as "means of joining, fastening" the various parts of a ship. In a slightly less direct way, in the Iliad, Homer uses the word harmony to describe an agreement or a covenant between people, *μάρτυροι ἔσσονται καὶ ἐπίσκοποι ἁρμονιάων,* "witnesses and guardians of our covenants" (Homer, 1925, p. 471). While the understanding of the word harmony in the Greek tradition evolved and expanded over time, and its meaning attained various attributes based on the context, it is important to emphasize three of these meanings due to their immediate relevance with the subject matter. These consist of:

A. Harmony personified as a divine figure

B. Harmony in music as a method of stringing, a musical scale, the octave, and as the planetary spheres in Pythagoreanism (Liddel, 2011, p. 244) and finally,

C. Harmony by way of "settled government" as the divine order of God Zeus (p. 244)

One of the earliest references alluding to the Harmony of the Spheres can be found in Aristotle's *Metaphysics* (Book XIV.vi.), associating the seven strings of the ancient heptachord with the constellation of the Pleiades: "There are seven vowels, seven strings to the scale, seven Pleiads; most animals (though not all) lose their teeth in the seventh year; and there were seven heroes who attacked Thebes" (Aristotle, 1935, p. 299). Nicomachus the Pythagorean mentions the doctrine in *The Manual of Harmonics*: "Among objects of perception, the music of the planets is considered to be the prototype of our music according as we imitate it" (Nicomachus, 1994, p. 45). However, the precise term *Harmony of the Spheres* appears for the first time in a fragment attributed to Empedocles found in the *Life of Pythagoras* by the Platonist philosopher Porphyry:

> He himself heard the harmony of the whole, because he had understood the universal harmony of the spheres of the heavenly bodies that move because of them—something that we other people do not hear because of the smallness of our nature. This is what Empedocles testifies to when he says about him. (Laks & Most, 2016 p.39.)

A Few Words on Orpheus' Seven String Lyre

In *The Lyre*, an Orphic book only known to us from commentary on Virgil, the scholiast links the seven strings of Orpheus' lyre to the seven planetary spheres: "

> But some say that Orpheus' lyre had seven strings corresponding to the seven circles of heaven. Varro says there was an Orphic book about summoning the soul, called the Lyre. It is said that souls need the cithara in order to ascend.

It is highly plausible that the book included a musical scale related to the planets and a description of the process Orpheus may have used with his lyre to assist the soul's ascension (West, 1983, p. 30).

Terpander, the Inheritor of Orpheus's Lyre

Prior to Pythagoras, there was Terpander of Antissa, a renowned musician and poet from the Greek island of Lesbos. One of his most noteworthy and enduring contributions is the advancement of the lyra from a four-string instrument to a seven-string one: "For you we shall make new hymns resound on a lyre of seven notes, abandoning our love for the four-voiced song" (Anacreon & Alcman, 1988, p. 319). Seven stringed instruments were already in use before Terpander's time by the Minoans (Kivilo, 2010, p. 117). The contributions of Terpander to the seven-string lyre should be understood as further harmonic innovations that were spread within the larger setting of the Ancient Greek music tradition.

Table 1: The Ancient Greek Heptachord and Terpander's Ancient Dorian Heptachord

(See Aristotle *Problems*, XIX, 32)

The Ancient Greek Heptachord	**Terpander's Dorian Heptachord**	**Corresponding Strings on the Ancient Greek Lyre**
D	E'	Nete
C	D	Paranete
Bb	C	Paramese or Trite
A	A	Mese
G	G	Lichanos
F	F	Parhypate
E	E	Hypate

In *The Fundamentals of Music,* Boethius states that Nicomachus not only attributes the addition of the seventh string to Terpander, but he also connects his innovation to the seven planetary spheres:

> In the beginning, Nicomachus reports, music was truly simple, since it was composed of four strings. It continued in this state until the time of Orpheus. In this period the disposition of strings was such that the first and fourth strings sounded the consonance of the diapason, while the mid-dle strings each in turn sounded the diapente and the diatessaron with the strings nearest them and those most distant. Indeed, there was nothing discordant in these, in imitation of cosmic music, which consists of the four elements. The inventor of the quadrichord is said to have been Mercury. At a later time, Toroebus, son of Atys and king of the Lydians added a fifth string. Hyagnis the Phrygian added a sixth string to these. Then a seventh string was added by Terpander of Lesbia, obviously in likeness to the seven planets. (Boethius, 1989, 29-31).

Knowledge on ancient Greek astronomy during the time of Terpander is very limited and therefore it is difficult to know for certain if Terpander associated his seven-tone scale with the seven planets or with the constellation Pleiades, as Aristotle hints in his *Metaphysics*. One thing can be said with confidence, Terpander's seven tone *harmonia* points to the stars. The story of Terpander as a great musical innovator goes well beyond the boundary of history, reaching the realm of legend and myth, and connecting him to Orpheus' own lyre: "While Terpander was still residing in Antissa, some fishermen found the lyre of Orpheus, which, along with his head, had floated to the northern coast of Lesbos after the Thracian was torn apart by enraged women, and gave it to Terpander" (Power, 2010, p. 2). Placing Terpander at the center of the myth and not only associating him with Orpheus, but also making him the sole inheritor of his musical instrument and tradition, gives us a valuable link between the music tradition established by Terpander and the Orphic mystery tradition.

The Invention of the Octave and its Attribution to Pythagoras

Using the musical innovations of Terpander as steppingstones, Pythagoras contributed his own improvements to the ancient Dorian *harmonia.* He separated the two conjunct tetrachords in the ancient seven tone scale (E'DCBbAGF) by adding a string between *mese* and *paramese* thereby creating two disjunct tetrachords (EDCB-AGFE) and thus completing the octave.

Table 2: The Ancient Octachord Attributed to Pythagoras and Based on Nicomachus' Description
(see Nicomachus, 1994, Ch. 5)

The Pythagorean Octachord	Corresponding Strings on the Ancient Greek Lyre
E'	Nete
D	Paranete
C	Trite
B	Paramese
A	Mese
G	Lichanos
F	Parhypate
E	Hypate

On Pythagoras' Ability to Hear the Harmony of the Spheres

In a fragment attributed to Embedocles, we learn that Pythagoras had the ability to hear the aforementioned "harmony of the whole" and that he was capable of that hearing because he "had understood the universal harmony of the spheres." Pythagoras had advanced his knowledge on theology, astronomy, and mathematics, and for him the Harmony of the Spheres was first and foremost an intellectual endeavor, and then a musical one. Like Terpander, Pythagoras is also associated with Orpheus and his mysteries. In a fragment referencing Ion of Chios, the dramatist and philosopher (who lived closer to the time of Pythagoras) claims that Pythagoras wrote poems and attribute them to Orpheus (see Pythagoras, *Testimonia*, Part C: Attributed Doctrines, D) (Laks & Most, 2016, p. 101). Other authors such as Iamblichus connect Pythagoras with Orpheus to the extent that they attribute all of Pythagoras' knowledge of the Gods to Orpheus himself (see Pythagoras, *Testimonia*, Part A: Person, P) (Laks & Most, 2016, p. 25).

Plato's Harmony of the Spheres

The first time that *harmonia* and astronomy are introduced as related sciences is in the seventh book of Plato's *Republic*, as transmitted by Socrates:

> ...that just as our eyes have fixed on astronomy, so our ears have fixed on harmonic motion and these sciences are related to each other, as the Pythagoreans claim, and we too agree, Glaucon. Or how do we see it? "As you've just said," he said (Plato & Emlyn-Jones, 2012, p. 165).

In this brief passage, Plato, using the words of Socrates, prepares the ground for the Harmony of the Spheres and announces that this is a Pythagorean notion. Later, Plato (in Book X of the *Republic*) closes his work with the Myth of Er (lines 614-621), a teleological myth that becomes the inspiration of an intense philosophical speculation among philosophers and across the ages. This is also the place in the *Republic* where Plato reveals for the first time the Orphic Pythagorean doctrine of the Harmony of the Spheres:

> And the spindle turned on the knees of Necessity, and up above on each of the rims of the circles a Siren stood, borne around in its revolution and uttering one sound, one note, and from all the eight there was the concord of a single harmony. And there were another three who sat round about at equal intervals, each one on her throne, the Fates, 1 daughters of Necessity, clad in white vestments with filleted heads, Lachesis, and Clotho, and Atropos, who sang in unison with the music of the Sirens, Lachesis singing the things that were, Clotho the things that are, and Atropos the things that are to be. And Clotho with the touch of her right hand helped to turn the outer circumference of the spindle, pausing from time to time. Atropos with her left hand in like manner helped to turn the inner circles, and Lachesis alternately with either hand lent a hand to each (Plato & Emlyn-Jones, 2012, p. 475).

Like his predecessors, Plato is influenced by Orphic theology and ideas. The fact that he replaces the traditional mother of the Fates, Themis (see Hesiod's *Theogony*), with the Orphic goddess Necessity, places the myth within the context of the Orphic tradition. Notions such as the soul's separation from the body, the doctrine of reincarnation, the waters of Lethe and Mnemosyne, the immortality of the soul, the soul's judgement in the afterlife, and the fate of the soul are all Orphic subjects that concern Plato regularly in the various myths he transmits in his dialogues, and the Myth of Er is no exception.

The Invisible *Harmonia* and the World Soul

With the Harmony of the Spheres included in the Myth of Er, Plato separates what he understands to be a divine and invisible *harmonia* of the Cosmos from a *harmonia* that is subject to the world of the senses, that is, our world. Consequently, Plato avoids giving a purely musical analysis and an obvious practical application to the Harmony of the Speres in the myth mainly for four reasons. Firstly, so he can separate the observable reality from the invisible divine world. Secondly, to transmit his notion that divine *harmonia* is superior to musical *harmonia*. Thirdly, that the second originates with the first, and finally, to accentuate the divine nature of numbers. The latter is a purely Pythagorean notion and adds another stratum to the myth, making it harder to attribute the doctrine only to Plato. The above-mentioned invisible *harmonia*, which is subject to esoteric and intellectual processes only, seems to be for Plato the only kind of *harmonia* that can lead to the prototypical and ideal world of his eternal immutable forms. As Iamblichus asks very elegantly in his *Life of Pythagoras* when he transmits the auditions of the *Acousmatici* philosophers: "What is harmony? That in which the Syrens subsist" (Taylor, 1986, p.43).

Plato is not the only philosopher to allude to an invisible and divine *harmonia.* Heracleitus, Empedocles, and the Pythagorean Philolaus also mention the hidden properties of *harmonia* and accentuate the importance of balancing opposing principles. Two of these fragments from Heracleitus stand out and are worth mentioning due to their relevance to the subject. As he states in fragment XLVII: "the invisible attunement is superior to the visible" (Hippocrates & Heracleitus, 1931 pg. 487). Similarly, in fragment XLV: "They understand not how that which is at variance with itself agrees with itself. There is attunement of opposite tensions, like that of the bow and of the harp" (pg. 485).

There may be a fifth reason that Plato does not introduce musical examples of divine *harmonia* in the Myth of Er*:* on one level, so he can veil it from the eyes of the profane, and on a second level, to make the pious seeker work with his intellect and imagination to experience it meaningfully. After all, Goddess Harmonia mentioned earlier is the daughter of God Zeus, the divine mind principle, an attribute of the heavens that Plato makes abundantly clear throughout his writings that he venerates. Therefore, it is only natural to assume that he will not give up the entire doctrine easily and that he will make the reader employ the power of the mind and imagination to better grasp the abstract qualities of an invisible *harmonia*. That leaves one with the question, where can this hidden *harmonia* in the myth be traced to, and how might it sound?

A few lines prior to the revelation of the Harmony of the Spheres, Plato gives, through the eyes of Er's soul, a series of perplexing planetary arrangements based on the rim width, color, light conditions, and speed of the whorls (planets) attached to the spindle of Necessity:

> Its shape was that of those in our world, but from his description we must conceive it to be as if in one great whorl, hollow and scooped out, there lay enclosed, right through, another like it but smaller, fitting into it as boxes that fit into one another, and in like manner another, a third, and a fourth, and four others, for there were eight of the whorls in all, lying within one another, showing their rims as circles from above and forming the continuous back of a single whorl about the shaft, which was driven home through the middle of the eighth. Now the first and outmost whorl had the broadest circular rim, that of the sixth was second, and third was that of the fourth, and fourth was that of the eighth, fifth that of the seventh, sixth that of the fifth, seventh that of the third, eighth that of the second; and that of the greatest was spangled, that of the seventh brightest, that of the eighth took its color from the seventh, which shone upon it. The colors of the second and fifth were like one another and more yellow than the two former. The third had the whitest color, and the fourth was of a slightly ruddy hue; the sixth was second in whiteness. The staff turned as a whole in a circle with the same movement, but within the whole as it revolved the seven inner circles revolved gently in the opposite direction to the whole, and of these seven the eighth moved most swiftly, and next and together with one another the seventh, sixth and fifth; and third in swiftness, as it appeared to them, moved the fourth with returns upon itself, and fourth the third and fifth the second. (Plato & Emlyn-Jones, 2012, p. 469-473).

Based on the above passage, the following planetary orders are ascertained:

A. Initial order of the eight whorls found one inside the other: 1 2 3 4 5 6 7 8
B. Order based on their rim's width: 1 6 4 8 7 5 3 2
C. Order based on their color and light conditions: 1 7 8 2 5 3 4 6
D. Order based on their relative speed: 1 8 7 6 5 4 3 2

Instead of using the names of the planets, Plato identifies them numerically based on their position, width, color, and speed in the cosmological model that he unveils in the Myth of Er. This is quite perplexing as Plato was not ignorant of the names of the seven planets, evident since he mentions them in one of his later dialogues, the *Timaeus.* So why does Plato withhold the names in the Myth of Er, choosing instead to give them numerical values? The first reason may be theological and the second, musical. Plato thinks like a speculative musician and more specifically, like a string instrument player who holds a lyre, which in this case is the invisible cosmic lyre of the Gods and Goddesses who use it to create the orderly arrangement of the invisible World Soul. As Plato mentions in the *Timaeus*: "And whereas the body of the Heaven is visible, the Soul is herself invisible but partakes in reasoning and in harmony." (Plato, 1929, p. 73). In this cryptic way, Plato reveals the existence of an invisible World Soul that partakes in number, reasoning, and *harmonia* nevertheless. The musical reasoning behind the numerical naming of the wandering stars is to bring the Harmony of the Spheres to life through number. Therefore, by combining knowledge of the ancient octachord attributed to Pythagoras and the planetary names as found in the *Timaeus (p. 79)* and using that information to view the four numerical arrangements as melodic sequences where numbers correspond to musical notes from the Dorian scale, the following musical patterns (approximated in the western notation system) can be ascertained:

A. Initial order of the eight whorls: 1 2 3 4 5 6 7 8
   Corresponding musical notes (ancient Pythagorean Dorian octachord): E' D C B A G F E

B. Order based on rim width: 1 6 4 8 7 5 3 2
   Corresponding musical notes: E' G B E F A C D

C. Order based on color and light conditions: 1 7 8 2 5 3 4 6
   Corresponding musical notes: E' F E D A C B G, and finally

D. Order based on relative speed: 1 8 7 6 5 4 3 2
   Corresponding musical notes: E' E F G A B C D

Table 3: Melodic Patterns Implied in the Myth of Er and Table of Correspondences

The Pythagorean Octachord (Western Notation)	Planets	Strings on the Ancient Greek Lyre	Original Order of the Eight Whorls (planets)	Order Based on Rim Width and Corresponding Note	Order Based on Color and Corresponding Note	Order Based on Relative Speed
E′	Fixed Stars	Nete	1	1 - E	1 - E	1 - E′
D	Phainon (Saturn)	Paranete	2	6 - G	7 - F	8 - E, 1st
C	Phaethon (Jupiter)	Trite	3	4 - B	8 – E′	7 - F, (second in speed)
B	Pyroeis (Mars)	Paramese	4	8-E′	2 - D	6 - G, (same as 7)
A	Stilbon (Mercury)	Mese	5	7 - F	5 - A	5 - A, (same as 6 and 7)
G	Eosphoros (Venus)	Lichanos	6	5 - A	3 - C	4 - B, (third in speed)
F	Helios (Sun)	Parhypate	7	3 - C	4 - B	3 - C, (fourth in speed)
E	Selene (Moon)	Hypate	8	2 - D	6 - G	2 - D, (fifth in speed)

If the above mentioned proposition about the numeric values of the eight whorls and their four arrangements correspond to the strings of a lyre, and consequently to their relevant musical tones, the following observations are made: The original descending order of the planets represents the journey of *harmonia* from the divine world starting with the realm of the fixed stars, then passing through the seven spheres, and finally arriving to the moon. The final arrangement clearly represents the opposite direction. The divine *harmonia* descends the whole octave from the fixed stars to the Moon and from there, the opposite journey begins – an ascent that will eventually end at its place of origin. The latter is in perfect alignment with the Greek theology regarding the role of the moon as a place where souls arrive or depart (Plutarch, 1959, p. 469-471). Additionally, in the numeric order based on the rim width of the whorls, two consecutive ascending triadic arpeggios appear: one Dorian (EGB) in nature, and one Hypolydian (FAC), separated by the note E (EGB, E, FAC, D). In Plato's Harmony of the Spheres synchronized movement, proportion, color, light, and the divine nature of numbers all work together to produce harmony and beauty in the divine world.

It is worth acknowledging that there are cosmological, astronomical, and other scientific implications of immense value throughout the Myth of Er. However, their exegeses and interpretations are beyond the scope of this work. Ultimately, by giving precedence to the numerical sequences of the planets, Plato directs attention to the fact that these sequences are somehow perceivable by the human soul in the afterlife and that they produce *harmonia*. In this way, Plato transmits veiled melodical patterns based on the Pythagorean octachord and ideas. After all, how else could Plato reveal a truly invisible and divine *harmonia* of cosmic proportions observable only by the mind's eye?

The Harmony of the Spheres in the *Timaeus*

Later in his life, Plato continues to evolve the doctrine while renegotiating the scientific knowledge of his time in a new cosmological model that he introduces in his dialogue, the *Timaeus*. The idea of *harmonia's* divine origin remains unaltered. However, *harmonia* becomes not as an invisible attribute of the divine mind, but an active agent contributing in the birth, formation, and equilibrium of the material World Soul. The planets in this new cosmological model are also material Gods whose primary purpose is to preserve time. Plato's Demiourge in the *Timaeus*, who is presented as a craftsman geometrician God, uses the four elements as primary substances to create the sensible Cosmos. In this primordial environment of creation and formation, the Demiourge employs geometric, arithmetic, and harmonic means for the creation of the World Soul, thus creating a perceivable material reality between man and the invisible divine realm infused with movement, harmony, and beauty in form.

A New Harmony of the Spheres System
An Original Heptatonic *Harmonia*

As I introduce my original Heptatonic *Harmonia* and Harmony of the Spheres system, it is imperative to clarify beforehand that as I approached this ancient subject harmonically, my main concern was to create a consonant backdrop that honors the harmonic approaches and aesthetic values of Terpander, Pythagoras, and Plato regardless of any apparent limitations, rather than simply create a stand-alone *harmonia* or scale for my system. The above-mentioned harmonic relationships are the ones implied or directly deriving from the arithmetic and harmonic relationships as found in the writings of the Pythagoreans, especially in the Pythagorean Tetraktys, Plato's *Republic*, and the *Psychogony* in the *Timaeus.*

Table 4: Table of Correspondences
The Original Heptachord: E' D C# B A F# E

Western Notation	The Seven Planets	Corresponding Strings on the Ancient Greek Lyre
E'	Phainon (Saturn)	Nete
D	Phaethon (Jupiter)	Paranete
C#	Pyroeis (Mars)	Paramese or trite
B	Stilbon (Mercury)	Mese
A	Eosphoros (Venus)	Lichanos
F#	Helios (Sun)	Parhypate
E	Selene (Moon)	Hypate

Intonation Systems and Instruments of Choice

The instruments used for composing my original music based on the various Harmony of the Spheres systems, including my own, are a scientifically reconstructed thirteen-string ancient Greek lyre and a six-string hollow body guitar. For the ancient Greek lyre, I selected the Pythagorean tuning system due to its production of more pleasant fourths, fifths, and octaves. For the guitar, I used the equal temperament system, with the A note being tuned to 440 Hz.

Uniting the Visible and the Invisible: A New Astronomic (Horoscopic) Approach to the Harmony of the Spheres

"Horoscopic Astronomy" is an original term I created to describe the act of using the relative position of the seven classical planets with special attention to their rising and setting times, hence the term *horoscopic*. Furthermore, I symbolically employ the planetary spheres in my original system as significators of their corresponding Greek Gods alongside the virtues and attributes which they embody in the environment of the ancient Orphic and Pythagorean traditions. In my original approach to the system, I use the term astronomy with its literary etymological sense as "the law of the stars" instead of the term astrology, "the logos of the stars." The reason being that the system honors:

A. The observed or calculated position of the planets as they rise, rotate, and set from any coordinates on Earth, regardless of their visibility, and
B. The ancient understanding of *harmonia* as the divine order of the Orphic God Protogonos/Pan/Zeus (Damascius, 2010, p.416).

Additionally, and unlike traditional Hellenistic Horoscopic Astrology which mainly aims to predict events based on the location of the planets among other objectives, my approach to the ancient and original system aims to position the person at an advantageous intellectual and moral ground via the art of correspondences such as planets corresponding to virtues and specific attributes of the Gods. Finally, Geocentricism is maintained throughout the system for the purpose of consistency with the ancient Greek tradition and is purely symbolic. In this way, and from a philosophical point of view, focus shifts to the person, humanity, and the mutable world of material reality, which is represented by the planet Earth, but only as an initial point of departure towards the starry sky.

An Original Harmony of the Spheres Composition and Two Musical Examples of Horoscopic Astronomy Data

The time, date, and locale of the seven classical planets used for the original composition and two musical examples below are set to June 19, 2021, Delphi (Peak) – Central Greece. The system for the examples below is adjusted to a lunar day which begins with the rising moon of June 18th, and it can be used equally well for both solar and lunar calendars.

Figure 1: Original Harmony of the Spheres Composition (Original Heptachord System)
Order Produced by the Planets: EE'DBF#AC#

To Apollo

Two Original Harmony of the Spheres Examples Utilizing the Ancient Systems

Figure 2: Example utilizing Terpander's Heptatonic Dorian system
Order Produced by the Planets: EE'DAFGC

Terpander's Heptatonic Dorian System

Note: Even though Terpander's cosmological model is unknown, I assigned the seven classical planets on his heptachord, similarly to Nicomachus, in order to bring his system to life and use it in the context of the Harmony of the Spheres system.

Figure 3: Example utilizing Pythagorean/Platonic system
Order Produced by the Planets: E'EDCAFGB

Pyhtagorean/Platonic Dorian Octachord

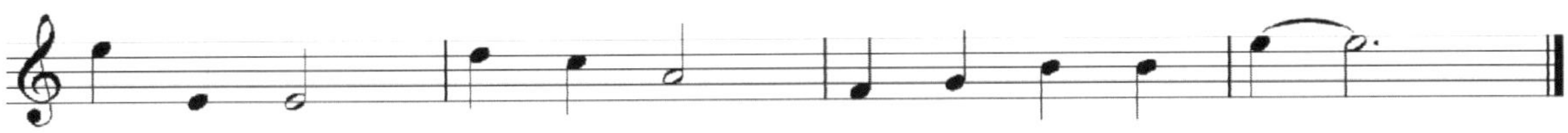

The Harmony of the Spheres as *Theon Therapeia*

In the ancient Greek world, especially during the time of Plato, the common understanding of the word *therapeia* (therapy) was broader than the understanding we have today, and it was primarily viewed as *service to the Gods* (TLG) as well as medical treatment for the human body. The implementation of the Harmony of the Spheres as *theon therapeia* is a noble pursuit, and at the same time, a challenging one since it demands one to partake in ritual acts of honest devotion in the context of Greek Religion and tradition. As it was alluded in the fragment related to Orpheus' seven string lyre, Orpheus and possibly those who followed him may have used this idea therapeutically in a ritual setting. In this case, the aim would have been to assist the soul to ascend via the seven spheres with the help of appropriate music sung and performed. Plato is very direct regarding *harmonia's* therapeutic effect on the human soul, and without any hesitation declares both the divine origin and restorative quality of musical harmony in the *Timaeus*:

> Music too, in so far as it uses audible sound, was bestowed for the sake of harmony. And harmony, which has motions akin to the revolutions of the Soul within us, was given by the Muses to him who makes intelligent use of the Muses, not as an aid to irrational pleasure, as is now supposed, but as an auxiliary to the inner revolution of the Soul, when it has lost its harmony, to assist in restoring it to order and concord with itself. And because of the unmodulated condition, deficient in grace, which exists in most of us, rhythm also was bestowed upon us to be our helper by the same deities and for the same ends. (Plato, 1929, p. 109).

In his poem "Hermes," Eratosthenes "brought Plato's planetary scale into connection with a divine lyre. He described how Hermes ascended to heaven and marveled to find the planets humming along their orbits on the very notes of the lyre he had invented on earth" (West, 1983, p. 30-31). The Harmony of the Spheres and the ascent of the soul are at the core of the doctrine and cannot be separated from their association with the Greek Gods since this is the environment out of which they come into existence.

A trained musician, by selecting a date, time, and locale to determine the rising and setting hours of the seven planets, can apply the Harmony of the Spheres system and adjust the music to the appropriate occasion. The use of the original system and approach in a ritual setting aims to help one form tangible connections with the heavenly dome as it rotates directly above.

Furthermore, it aims to seal the participants in an environment of piety and devotion, free from the profanities of the outside world, and to help their souls achieve the desirable ascent by invoking higher musical aesthetic values based on consonant sounds. Finally, via the art of correspondences, the seven planetary spheres become significators of the Gods and the virtues they are associated with, providing the means for a higher moral ground via inspiration.

The Harmony of the Spheres and Moral Theory in Music Therapy Practices

During Plato's time, when sophists were concerned with the teaching of virtue, music learning was the law. In the *Republic*, Plato places particular emphasis on the ethos of musical modes and rhythms and discusses how they are related to human behavior and virtue within the state. In the dialogue between Socrates and Glaucon, Socrates discusses the ethos and usage of musical modes and rhythms. Specifically, Socrates makes the point that out of all the musical modes, there are only two that are of value, the Dorian and the Phrygian (Plato & Emlyn-Jones, 2013, p. 273-274). Later in the dialogue, Socrates discusses the subject of rhythm and melody: "rhythm and melody above all penetrate to the innermost part of the soul and most powerfully affect it, bringing gracefulness, and, if one is brought up correctly, make one graceful; if not, isn't the result the opposite?" (Plato & Emlyn-Jones, 2013, p. 283). Through Socrates's words, Plato aligns himself with sophist traditions while highlighting the importance of music in shaping one's character, and how that character may affect the person and the state. According to the philosopher, even though these ideas concern an ideal state, they still have a high degree of value.

Today the music therapist has a lot to gain from this highly evolved moralistic approach to music. By identifying the type of morale and level of stimulation they want to use on the patient, and by applying the appropriate ancient modes, genera, and rhythms with their respective ethos, the practitioner can create an appropriate environment for the patient and begin the process of healing.

A practitioner with basic knowledge of music composition can create simple songs with high therapeutic value since all three Harmony of the Spheres systems proposed here are inherently consonant. The melodies and their implied harmonies can be produced using the techniques introduced earlier with the added step of adjusting the newly composed music to the appropriate rhythms for the patient's needs. For further optimization, one can refer to the ethos of each ancient Greek mode, as described by the ancient Greek authors and commentators on ancient Greek music.

Conclusion

Even though direct evidence regarding the Harmony of the Spheres doctrine is scarce, the ancient musicians, writers, and philosophers left enough clues for us to admire the beauty of the invisible and divine *harmonia* in the works of the Demiourge. While challenging our perceptions with their oracular words, they invite us across the ages to take an esoteric journey, expand the limits of our imagination and intellect, and elevate our souls to a higher intellectual and moral ground. The effort to bring this system to life is a worthy one in that it creates a tangible continuity with the ancient past, honoring all those who created, preserved, and passed it down. Moreover, it is a practical system: an ascension engine of tremendous potentiality and not merely a theoretical construct for intellectualization. When used correctly in a therapeutic setting, it can become a powerful agent of restoration, transformation, and progress.

References

Anacreon, & Alcman. (1988). Greek lyric: Anacreon, Anacreontea. Loeb Classical Library.

Aristotle. (1935). *Metaphysics, Volume II: Books X-XIV* (H. Tredennick, & G. C. Armstrong, Trans.). Loeb Classical Library.

Aristotle. (2011). Problems, volume I: Books 1-19. Harvard University Press.

Athanassakis, A. N., & Wolkow, B. M. (Trans.). (2013). The orphic hymns. Johns Hopkins University Press.

Boethius, Bower, C. M., & Palisca, C. V. (1989). Fundamentals of music. Yale University Press.

Damaskios. (2010). Damascius' problems and solutions concerning first principles (S. Ahbel-Rappe, Trans.). Oxford University Press.

Hesiod. (2018). Theogony - Works and Days - Testimonia (G. W. Most, Trans.). G. W. Most (Ed.). Loeb Classical Library.

Hippocrates, & Heracleitus. (1931). Heracleitus: On the universe (W. H. Jones, Trans.). Loeb Classical Library.

Homer. (1919). *Odyssey, volume I: Books 1-12* (A. T. Murray, & G. E. Dimock, Trans.). Loeb Classical Library.

Homer. (1924). *Iliad: Books 1-12* (A. T. Murray, & W. F. Wyatt, Trans.) (p. 13). Loeb Classical Library.

Homer. (1925). *Iliad: Books 13-24* (A. T. Murray, & W. F. Wyatt, Trans.) (p. 13). Loeb Classical Library.

Homer. (2003). *Homeric hymns, homeric apocrypha, lives of Homer*. M. L. West (Ed.). Loeb Classical Library.

Kivilo, M. (2010). Early Greek poets' lives: The shaping of the tradition. Brill Academic Pub.

Laks, A., & Most, G. W. (2016). Early Greek philosophy: Western Greek thinkers, Part 1.

Liddell, H. G., Scott, R., & Jones, H. S. (2011). μουσική. In the online Liddell-Scott-Jones Greek-English lexicon (p. 1148). http://stephanus.tlg.uci.edu/lsj/#eid=71140

Liddell, H. G., Scott, R., & Jones, H. S. (2011). ἁρμονία. In the online Liddell-Scott-Jones Greek-English lexicon (p. 244). http://stephanus.tlg.uci.edu/lsj/#eid=15652

Nicomachus. (1994). The Manual of harmonics of Nicomachus the pythagorean. Red Wheel/Weiser.

Plato. (1929). Timaeus ; Critias ; Cleitophon ; Menexenus ; Epistles (R. G. Bury, Trans.). Loeb Classical Library.

Plato, & Emlyn-Jones, C. (2012). Republic, volume II: Books 6-10. Loeb Classical Library.

Plato, & Emlyn-Jones, C. (2013). Republic, volume I: Books 1-15. Loeb Classical Library.

Plutarch. (1959). Moralia: On the love of wealth; On compliancy; On envy and hate; On praising oneself inoffensively; On the delays of the divine vengeance; On fate; On the sign of Socrates; On exile; Consolation to his wife (P. H. Lacy, Trans.). Loeb Classical Library

Power, T. C. (2010). The culture of Kitharôidia. Harvard University Center for Hellenic.

Siculus, D. (1939). *Library of history* (C. H. Oldfather, Trans.). Loeb Classical Library.

Taylor, T. (1986). Iamblichus' life of Pythagoras. Simon & Schuster.

West, M. L. (1983). The orphic poems. Oxford University Press.

On Music and Grief: Using Song to Help a Community in Trauma

Jonathan Seligman

Abstract

The present article is a narrative over a study. After the passing of an important 6th grade teacher, music director Jonathan Seligman teaches his school choir the song "Saturn" by Sleeping At Last to not only commemorate the loved one but to help the community through the journey of grief. This narrative shares the aftereffects of the teacher's passing, how the school temporarily coped with grief, the process of learning the song, and impact of the song for the choir and community abroad.

Keywords: music education; trauma; grief; song; choir; Sleeping At Last

Note: This article was first featured in the American Center for Elemental Music and Movement Publication on January 21, 2021. It is reprinted here with their permission.

Introduction

As music teachers, we thrive off the moments that touch a student's life. There is excitement when a shy student finds their confidence on the violin; excitement when an English Language Learner finds connection in choir. These moments of impact affect us for the rest of our lives. Last year, I had that same opportunity to create impact—but it was not out of joy, rather out of tragedy and necessity. My mentor and friend, our school's beloved 6th grade teacher, Lee Sison, unexpectedly passed away.

Mr. Sison was a staple of our community, teaching at Casillas Elementary for 16 years. He was more than "just a teacher". Within his time at Casillas, he fostered a culture of family at the school. With Lee, to be a student was to be more than an individual who learned facts. You were a part of something bigger. Lee helped you feel that—even if you weren't a part of his class. He bestowed that same feeling amongst his coworkers. Lee welcomed you with open arms and placed you in positions to grow the community and yourself. So when he passed away, there was a void. A pain and a void.

A Song to Reconcile the Void.

The community at large gathered together. Parents of former students led a vigil right outside the school. The community shared their pain and sorrow. Many were there during Lee's funeral to commemorate his life and celebrate his impact.

After the funeral had passed, life pushed forward. Quarter 2 came along and a new teacher took over his classroom. From the shadows, the teachers grieved, yet on the surface, we pushed forward. We had to for our students.

But our students weren't okay—especially those who had him as a teacher. At the age of 11, how does one process your leader being gone? How does one wrestle with the man that you looked up to no longer being present in your life? How do you even ask for help? Asking an adult for wisdom isn't going to bring back the man you miss. For some, to shed tears is to show weakness. You're in 6th grade. You're almost in middle school—two steps shy of adulthood. You're "supposed" to have this figured out. So how do you ask for help? How do you find closure with unexpected losses? Even at the age of 29, I still ask myself that question.

About a month after Lee's passing, I came across the song "Saturn" by Sleeping at Last. The piece starts out with a long string exposition, full of tension and decay, layers and minimalism. The tension grows until 1:55 when the voices finally enter on a melismatic "ooh," only to be decayed by a string unison and the dying of the beat of the toms. Then come in the words.

> You taught me the courage of stars before you left
> How life carries on endlessly even after death
> With shortness of breath you explained the infinite
> How rare and beautiful it is to even exist
>
> I couldn't help but ask you to say it all again
> I tried to write it down, but I could never find pen
> I'd give anything to hear you say it one more time
> That the universe was made just to be seen by my eyes
>
> With shortness of breath I'll explain the infinite
> How rare and beautiful it truly it is that we exist

This was the song we needed to sing. Despite it being a terrible choice pedagogically—being over an octave and a half wide in range, bearing odd syncopation, residing in the chest voice—we needed to sing this song.

To the non-religious, this song was a safe space to send out their turmoil and anger. For the religious, this song was an opening to believe once again despite the tragedy in their life. This was a song for all to draw out of the shadows the pain and emotions we felt and bring to the light that we were all still suffering. To sing this song as a choir would healthily acknowledge that we suffer together. We all miss Mr. Sison, but we miss him together. That despite the fact that the powers of time still push the world along, we still grieve.

And that's okay.

Rehearsing and Performing a Song of Hope.

Bringing this song to the students was not easy. From 4th-6th grade, after showing them the song in January, the students were stunned at the poignancy of the words to their own emotions, giving a true emotional response without reservations. Five months later, they all still deeply missed Mr. Sison. Many shared their hurt. Some questioned why we were doing this song at all. But after the many thoughts, they unanimously agreed that this song needed to be sung.

From thereon, choir became more than just a time to sing. We gave space to have students share memories of their late teacher. Former students would even walk in and share their stories as well.

In conjunction to the singing, the choir did a movement activity while the string players—all former students of Casillas—played the opening. The choristers had light-up orbs during the instrumental section to move around, representing the memories they had of Mr. Sison. Moving in improvisation, the orbs conjoined and separated. Once the lyrics began, students held onto their orbs under their faces, lit in fluorescent blue and yellow. Lit like candles.

The concert was not easy. "Saturn" being the penultimate song, there was a great mix of emotions afterwards. Some shocked, but most were extremely positive. Lee's family had been to many events that commemorated him, but they admitted that this was uniquely special. Strong-willed students who did not shed a tear throughout the year were red-eyed. Colleagues were thankful. There was a sense of release that some had not the ability to express.

What is Our Role During Times of Grief?

As music enthusiasts, we understand the power of the auditory arts. As music educators, we advocate for what we teach because we know music can heal. A song can bring happiness to a child who is having problems at home. A dance can bring joy to a child who is having trouble connecting with others.

But grief? Grief is hard.

There is no word that can bring a child out of grief as we call will continue to grieve after loss. Even for adults, grief is hard. We try to compartmentalize the pain and grope in the darkness for the good in the world. We try to ignore it. We dig it in a hole and box it away. And that often leads to isolation.

But what if we leaned into it? What if we taught our students to lean into it? You'd be surprised what you would find.

> "She told me that couldn't help but cry [when singing "Saturn"] but that they were tears of healing."
>
> -Luisa, mother of a 5th grade chorister

> "It was hard for me to finish the song without tears. Every word I sang, every beat, a memory of me and [Mr. Sison] flashed before my eyes. It reminded me of the pain and hurt me and all the other kids in his class had experienced earlier this year. [It reminded me of] how we grew and overcame that sorrow."
>
> -Lilly, 6th grade chorister

> "Music has always had a way of tugging at my heart strings, and if it's kids singing, it makes it even more emotional for me. Add in that my own son was singing, as well as many of my students, and it made this performance an extremely emotional experience. Listening to this performance brought up many emotions very quickly. We as a community went through so much last year with the loss of our beloved Mr. Sison, and I feel like some of the kids didn't get a chance to really grieve and face their emotions during this tragic time. I felt that this performance was a necessary release of emotions for our whole community. There wasn't a dry eye in that room, and it brought us closer together and I think it added to our healing process. "
>
> -Melissa, Teacher/Parent and Friend of Lee

> "What courage it takes for these young ones to rehearse and practice and perform such a beautiful tribute! Not only did their performance help each other but helped the community as well. The power of music, in this performance is evident and much needed as everyone grieves in their own personal way. I believe and admire these children, under Jonathan's direction, to begin the healing process and "moving on" through their music and song. Somehow and in some way, I know my brother's spirit was present on that night, and just as he was in life, always cheering and supporting his kids."
>
> -Donna, sister of Lee

There is no one right way to handle loss. Grief is a pain that no doctor could remedy. But through vulnerability, there is empathy. Through empathy, there is understanding. Through understanding, maybe there could be healing. Vulnerability can't be taught in a textbook. But through song, we have the power to invite the victim in, to suffer and heal in community.

References

Seligman, J. (2021). On Music and Grief: Using Song to Help a Community in Trauma. Reprinted with permission from https://acemm.org/2021/01/on-music-and-grief-using-song-to-help-a-community-in-trauma/

Self and Communal Healing by Vocalizing Chakra Toning Sounds While Dancing in Raves and Parties

Rona Geffen

Abstract

While dancing in raves and parties people tend to spontaneously vocalize vowels which accurately resemble what is commonly known as Chakra Toning Sounds. This sonic observation raises the hypothesis that people instinctively and intuitively heal themselves by balancing and activating their own chakras with their voice. To test and verify this hypothesis a survey was conducted in which 96 participants were asked to share their experience about the voices they produce and hear in raves and parties while dancing. The majority of participants reported using their voices and hearing others use their voices with Chakra Toning Sounds. An interesting observation suggests that a communal effect is created by this action, in which self-healing informs and encourages communal (group) healing, possibly underlying collaborative and collective healing.

Key Words: Vibrational Medicine, Chakra Toning Sounds, Dance, Rave, Self-Healing, Group Healing, Sound Healing,

Introduction

Raves of electronic music, festivals, popular music concerts and parties draw large groups of people who listen and dance to music together. The music usually contains repetitive rhythms and beats and is typically introduced using powerful sound systems. This study explores a hypothesis suggesting such events hold a strong relation to self-healing by the use of voice and Chakra Toning Sounds (CTS), which inform and encourage communal (group) healing.

Scientific evidence from ancient cultures demonstrates that from the beginning of humankind people have gathered to listen to music, dance and sing together (Bundo, 2001; Clarke, et al., 2015; Clayton, et al., 2004; Redmond, 1997). The use of voice in this context is traditionally referred to as primordial and tribal, with emphasis on singing. This study provided evidence that the use of voice in musical gatherings could in fact be considered as a form of sound medicine, a class of vibrational medicine.

Sound medicine is an ancient practice used in different cultures around the world (Ferguson, N.D.; Nunner, 2017; Reid, 2010) to relieve physical pain and emotional turmoil through the use of sound and rhythm (Franka, et al., 2000; Jingjing, 2015). Sound medicine incorporates the use of various instruments, including gongs, singing bowls, tuning forks, drums, and the human voice. The natural mechanisms on which sound medicine relies are entrainment and coherence of frequencies and rhythms. These are used to ignite physiological responses such as enhanced hormone release, for example Oxytocin (Uvnäs-Moberg & Petersson, 2005) adjustment and entrainment of brain waves according to vibrational intervals, also known as binaural beats (Lane, et al., 1998; Oster, 1973) and induction of stress and pain relief mechanisms (Gutgsell, et al., 2013; Kasprzak, 2011; Lee, et al., 2012). There are various traditions and practices for using voice in sound medicine from breathing to humming, mantras and chanting, all of which incorporate the vagus nerve, the tenth cranial nerve and the longest nerve of the autonomic nervous system, via a well-defined neurological tract.

Within sound medicine there are a few practices to activate one's chakras with sound and voice, for example the Bija mantras, the chakras related notes and CTS (Figure 1). The seven CTS are different vowels, each connected to a different chakra. Keyes(1973), one of the earliest pioneers in vocal toning and CTS, refers to toning and vocalizing sounds as a freeing and releasing practice. According to D'Angelo (2012) the vowels are primary sounds which allow richness of harmonics. Hausauer (2008) related to toning CTS as using the voice specifically as a healing instrument, stating that creating a drawn-out vowel or syllable using the voice can help restore balance and flow to the chakra system.

CHAKRA	CTS	BIJA MANTRA	NOTE	COLOR
Crown	EEE	OUM (or SILENCE)	B	Purple or White
Third Eye	AYE	OUM (OHM)	A	Indigo
Throat	EY	HAM	G	Blue
Heart	AH	YAN	F	Green
Solar Plexus	OH	RAM	E	Yellow
Sacral	OOH / OOO	VAM	D	Orange
Root	HUH / UH	LAM	C	Red

Figure 1: Chakras related Chakra Toning Sound (CTS), Bija Mantras, Notes and Colours

Chakra is a Sanskrit term which means “wheel” or “disk.” Today it is common to relate to seven chakras (Root, Sacral, Solar Plexus, Heart, Throat, Third Eye, Crown), though different practices suggest there are 4, 5, 6, 8 or more chakras. Chakras are considered to be energy centres of the body by many eastern medicine practices. In many integrative medicine practices chakras are considered to be highly influential and significant for spiritual, emotional, and physical balance and coherency. The physical location of most chakras coincides with the endocrine system and major nervous plexi, and thus could be related to hormonal activity in the human body and its related nervous system activity (Figure 2). Anodea (1987) points out the similarity between the positions of the chakras and the endocrine system as well as their related function and purpose.

This study demonstrates that the spontaneous act of self-vocalizing while dancing in raves and parties directly correlates to vowels and sounds that accurately resemble CTS in a way that seems to be instinctive and intuitive self-healing. Therefore, this study suggests raves and parties hold a strong relation to self-healing and deep inner vibratory mechanisms. An interesting observation of this study suggests a communal effect created by this action, where self-healing by the use of voice and CTS informs and encourages communal healing, possibly underlying collaborative and collective healing.

It should be noted that while people also tend to use their voice loudly at live performances, the current study investigated spontaneous vocalization without the mediation of a performer, a different case in which sounds are produced in the realms of encouragement and affirmation, rather than pure self-expression. This study shows that the type of sounds people use spontaneously are in line with the seven CTS system.

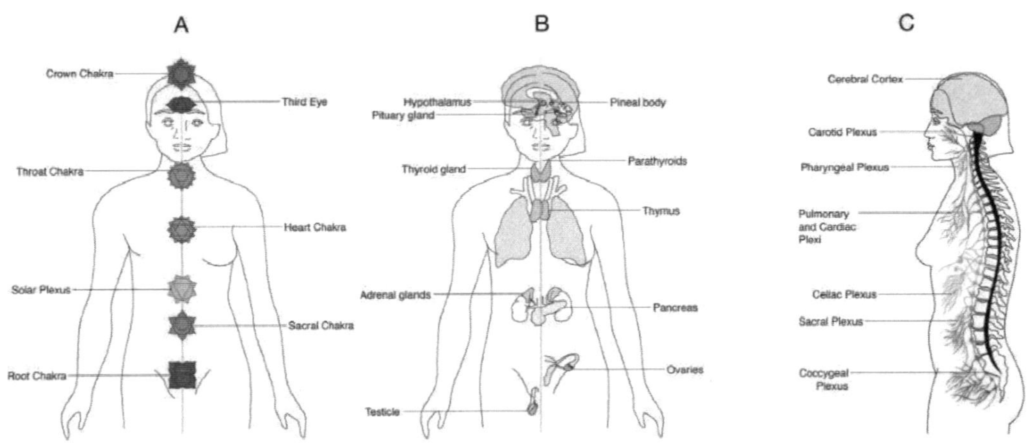

Figure 2: (A) The 7 chakra system positioning (B) Endocrine glands positioning (C) Illustrative sketch of the nervous plexi positioning

Methods

An overall 96 study participants were asked to fill in a questionnaire regarding their experience in audible vocalized sound during raves and mass dance parties. First, a group of 48 participants at O.Z.O.R.A. festival (summer 2019) was approached during the festival to fill in the questionnaire, with technical assistance of the researcher (Q1). Next, the questionnaire was mildly adjusted and uploaded online for a period of two months, in which additional 48 participants filled in the questionnaire (Q2). Participants were asked to describe the sounds they produce with their own voice, as well as sounds they hear other people produce with their voice while dancing in raves and parties (for the full questionnaires please see Appendix 1: tiny.cc/9252vz & 2: tiny.cc/c252vz). Responses to the two questionnaires were analyzed separately. The average and overall results of responses to both questionnaires were also computed. In addition, interviews were conducted with specialists in the fields of vibrational medicine, singing and voice research.

Results

This research includes 96 participants, all under 70 years old. Most participants were in the age groups 20-40 (36.5% are 20-30, 35.4% are 30-40). It is interesting to note that though ravers and party goers are usually considered to be in their twenties or even younger, approximately 60% of participants are over 30 years old, and 25% are over 40 years old.

A computed average over both questionnaires showed that 92.7% (Q1:93.8%, Q2:91.6%) reported to use their voice and 93.8% (Q1:93.8%, Q2:93.8) reported hearing others use their voice while dancing. This demonstrates that the absolute majority of participants report experiencing either using their own voice or hearing others use their voice while dancing in raves and parties.

In both produced and heard sounds, participants were asked which are most prominent on the dance floor. Participants were given multiple choice selection (options as described in Figure 1: HUH / UH, OOH / OOO, OH, AH, EY, AYE, EEE and OTHER). CTS were largely referred to in both produced and heard sounds on the dance floor. Among 89 participants who reported to use their voice while dancing (Q1:45, Q2:44) and given the multiple choice, a total of 196 responses (Q1:91, Q2:105) were collected to the question "Describe the type of sound/s you make while dancing." Out of the 196 overall responses, 153 of responses (Q1:71, Q2:82), reported using CTS and 43 of responses (Q1:20, Q2:23) reported using other sounds (Figure 3A). This result indicates that 78% of responses actually report using CTS while dancing.

A total of 230 responses (Q1:107, Q2:123) were collected to the question "Describe the type of sounds you hear other people make while dancing." Out of the 230 overall responses,

203 of responses-(Q1:95, Q2:108) reported hearing CTS, and 27 of responses (Q1:12, Q2:15) reported hearing other sounds (Figure 3B). This result indicates that 88.2% of responses report hearing other people using CTS while dancing. A closer observation of responses reveals that in fact some of the sounds participants describe as "other" actually resemble CTS. For example, howling can be related to HUH, Root chakra sound; or WHA-WHA can be related to AH, Heart chakra sound. In this case a new computation was calculated in which 165 responses (Q1:77, Q2:88) reported using sounds related to CTS and 32 responses (Q1:14, Q2:18) reported using other sounds (Figure 4A). Further, 211 of responses (Q1:100, Q2:111) reported hearing sounds related to CTS and 15 responses (Q1:7, Q2:8) reported hearing other sounds (Figure 4B). From now on the latter computation, including the related "other" sounds, will be further addressed. According to this computation, 84.1% of responses reported using CTS and 91.7% of responses reported hearing CTS on the dance floor. Between all types of computations, responses referring to CTS are largely more prominent, ranging from approximately 3 to 14 times more than using and hearing other sounds. It is interesting to note that while people report using "other" sounds relatively often, they also report hearing these sounds at the lowest likelihood on the dancefloor (Figure 4A, 4B).

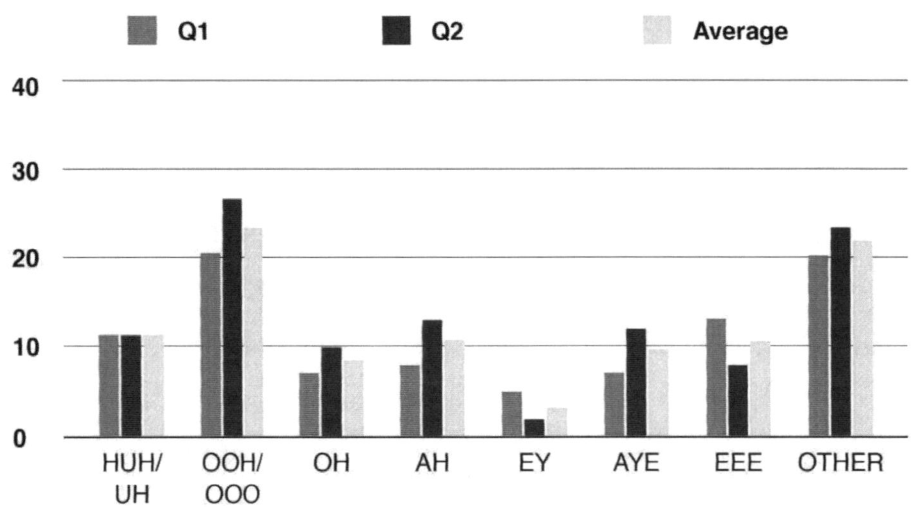

Figure 3A: The type of sounds people reported to produce on the dancefloor (Question in Questionnaire: Describe the type of sound/s you make while dancing)

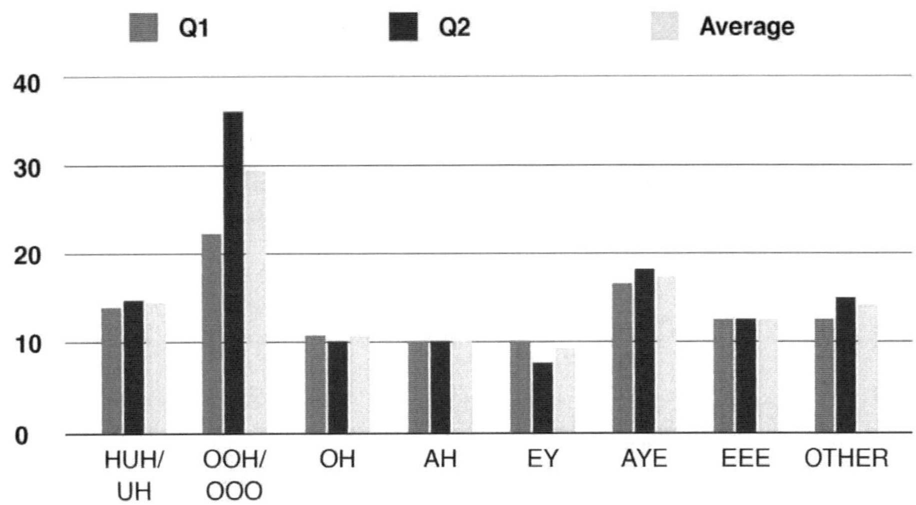

Figure 3B: The type of sounds people reported to hear on the dancefloor (Question in Questionnaire: Describe the type of sound/s you hear other people make while dancing)

It should be emphasized that even though participants had an option to choose multiple answers, it was evident that in both questionnaires the most prominent sound people reported to produce (47 of all responses, 23.9%. Q1:20, Q2:27. Figure 4A), and hear (61 of all responses, 26.5%. Q1:25, Q2:36. Figure 4B) on the dance floor is OOH/OOO, which is the Sacral CTS. Additionally, both questionnaires revealed that the EY sound (Throat chakra) is reported to be the least in use while dancing (Q1:5, Q2:2. Figure 4A). However, reports relating to what sounds people hear show similar numbers among both questionnaires between the sounds OH, AH and EY CTS accordingly (Solar Plexus: Q1:11, Q2:10; Heart: Q1:10, Q2:10; Throat Q1:10, Q2:8. Figure 4B).

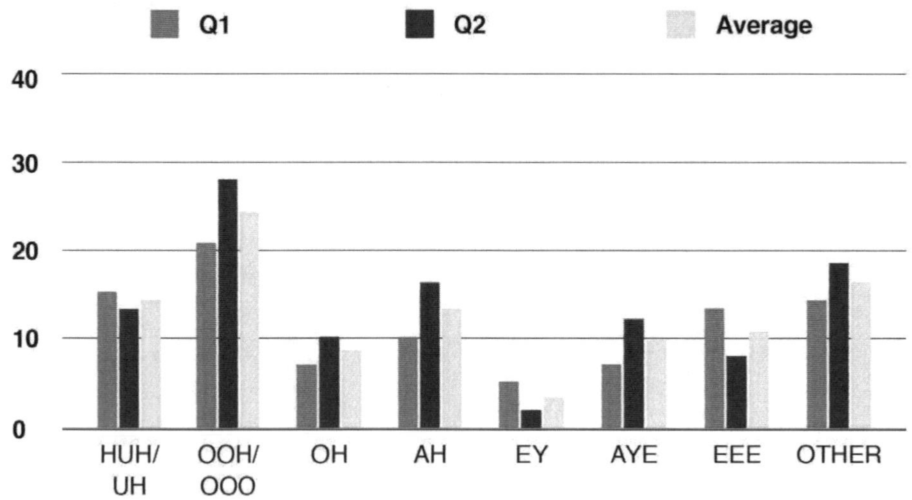

Figure 4A: The type of sounds people reported to produce on the dancefloor including relevant "other" sounds

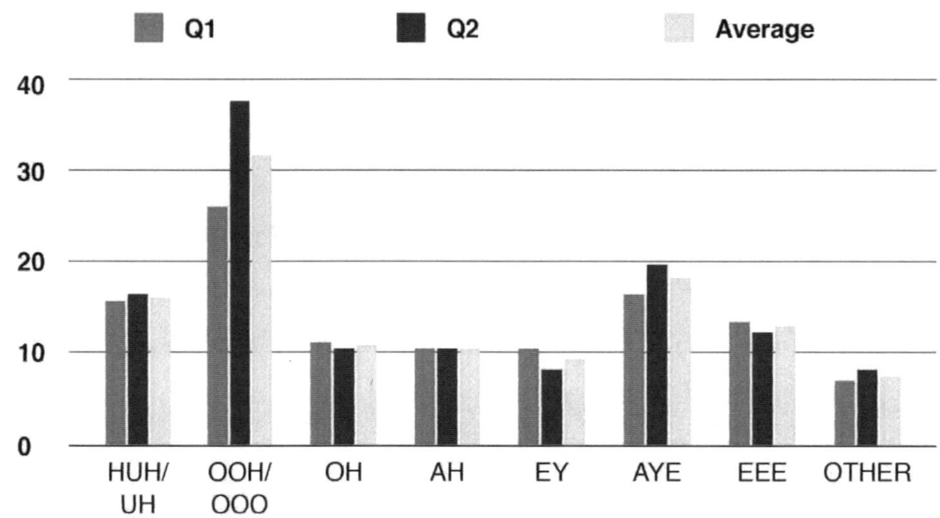

Figure 4B: The type of sounds people reported to hear on the dancefloor including relevant "other" sounds

When people were asked what makes them use their voice the way they do while dancing, three typical response types came up:

1. Music Related - A rhythmic, melodic, or harmonic event (or any combination of the three) was driving them for a more enhanced expression other than dancing. Responders tended to address bass and rhythm in particular.

2. Expressing Euphoric Feeling - Inner feeling was so euphoric or intense they no longer could express it only by dancing.

3. Answers that related to both these elements as one were also common (Figure 5).

Here are a few quotes from participants to the question "What makes you use your voice in this way while you dance?" This question was not mandatory and was answered by 79 (Q1:37, Q2:42) people. Examples for music related responses include "The more powerful the beat gets," "A certain note, drop, hit in the music that makes me feel like it." Examples of euphoric feeling responses include, "To express loudly my energy and my happiness," "Energy from within that can't express itself only through dance & motion & smiles & hugs, so I need to scream :)." Examples for both music and euphoric feeling responses included: "I'm connected with the music in a different kind of way, just feels like there are different channels for the music to come out of me again as it goes in my body and moves my soul and everything," "The music which is creating a very high energy in me and I need to take this energy out" (For a full list of all quotes please see Appendix 3: tiny.cc/e252vz).

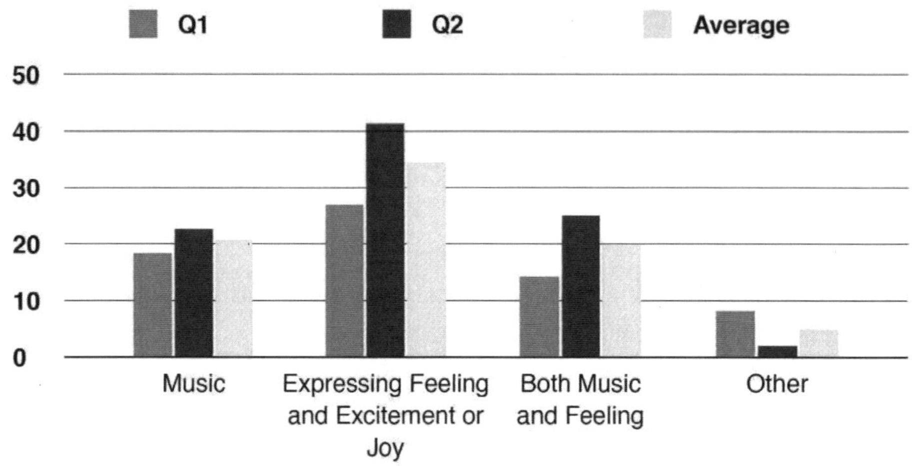

Figure 5: The reasons why people are using their voice while dancing (Question in Questionnaire: What makes you use your voice while you dance?)

When trying to break down the patterns, reasons, and sources for the use of each sound, responses were also not mandatory. This section tends to be more personal and under the conditions this research was conducted, an effort was made to minimize any relation to judgment or obligatory personal self-exposure. Subsequently, 59 of the 96 participants (Q1:23, Q2:36) answered the question relating to consistency in oral expression on the dance floor throughout the years. Out of these 59 participants, 32 (54.2%) reported that the sounds they produce while they dance have been consistent throughout the years. An interesting observation of the participants who reported the sounds they produced while they danced have changed over the years and some of the quotes from people who reported a consistent form of oral expression, is the relation they make with how comfortable they feel. It implies that the more comfortable people feel with themselves and others the more they allow themselves to self-express in various and more instinctive and uncommon ways: "(I'm using) More sounds as I'm getting older. Probably because I feel more comfortable", "I used to be too shy and anxious about how I sounded so I would just hum sometimes to the music" (For a full list of quotes please see Appendix 4: tiny.cc/g252vz).

The relation between the type of sounds people produce and what they are dealing with - physically or emotionally was also addressed. This question was not mandatory as well, and many participants chose to keep this confidential. Nevertheless, some interesting insight can be raised from the people who did answer (64 people overall, Q1:28, Q2:36). Responses suggested an indication between the type of sound used and the type of issue people are dealing with. For example, of the 28 people who reported using the sound AH (Heart Chakra), 16 answered this question. Notably, 75% of them reported dealing with either upper body physical issues, emotional issues, or both. Another example shows only 10 people reported lower body (waist down) physical issues, 80% of them reported using OOO and/or HUH sounds (Root and Sacral chakras - the lower chakras). These examples show correlation between the type of CTS used and the physical and emotional issues people are dealing with. However, as this question was not mandatory there is not sufficient data to compare absolute percentages and conclude a definite connection. This aspect of the research should be further explored in future studies.

Another interesting key observation is the communal aspect created by this action of self-vocalization in a group. Evidently people are encouraged to use their voice more when they hear others do the same. On average 67.8% of participants reported they are more than moderately encouraged to use their own voice when they hear others do so on the dance floor (Figure 6).

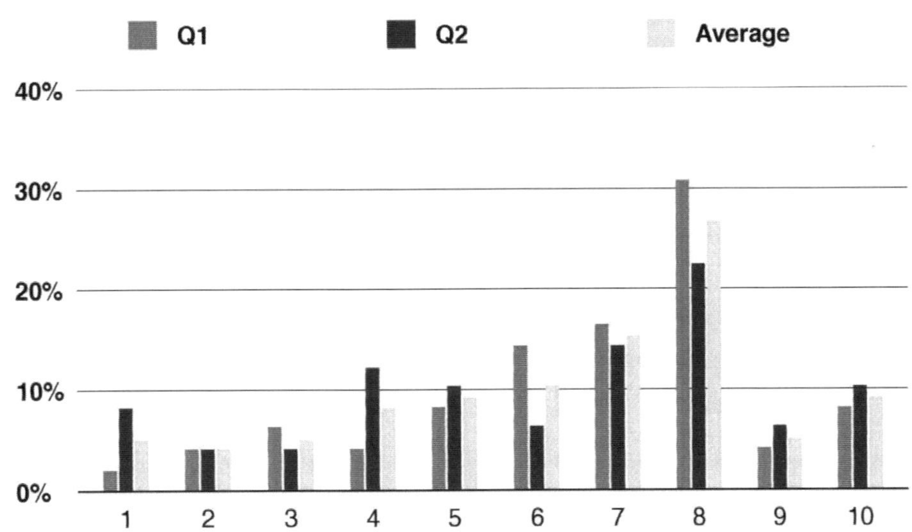

Figure 6: Communal effect created by self-vocalizing while dancing (Question in Questionnaire: Between 1 to 10 - To what extent do you feel encouraged in your own vocalization by the vocalization of the people around you?)

Discussion

The findings of this study suggest that people who listen to music and dance together in raves and parties, intuitively use their voice with CTS to heal themselves by balancing and activating their chakras, an action which seems to create an effect of group healing among participants. Being a social species, such integration of self with its community and their immediate influence underlies the importance of human interaction and its connection to wellbeing.

Raves of electronic music, festivals, and parties draw large groups of people who listen and dance to music together. Schamis (1985) denoted 8 mechanisms taking place while engaging in group dance therapy, which she related to as healing processes: Synchrony, Expression, Rhythm, Vitalization, Integration, Cohesion, Education and Symbolism. These processes refer both to individual and group healing mechanisms while one is involved in Dance Therapy, hence, dancing in a group. In the context of dance the use of voice is usually referred to as primordial and tribal with emphasis on singing. The findings of this study suggest it also holds a strong relation to self-healing and deep inner vibratory mechanisms, much like cats heal themselves by purring (Von Muggenthaler, 2001). These processes are attributable to both individual healing, as well as to group healing mechanisms which one is involved in while dancing in a group. It is well known that music has the power to relieve stress and anxiety (Harada, et al., 2017; Shin & Kim, 2011) and affects our mood (Schäfer, et al., 2013; Wells & Hakanen, 1991), energy and physical state (Gutgsell, et al., 2013; Lee, 2016). Clarke, DeNora, & Vuoskoski (2015) stated:

> Music is a powerfully multi-sensory, and particularly kinesthetic phenomenon whose embodied character draws people into fluid and powerful social groups at a range of scales and degrees of (im)permanence, and in doing so helps to enact a kind of empathy (p.19).

Coherence and synchronization taking place while collectively engaging in physical activity, specifically of a rhythmic order, is known to boost performance, release endorphins, forge friendships, and strengthen relations within participants (Taylor, Cohen, Davis, & Weintraub, 2016). This nicely resonates with theories supporting the notion that synchronized physical activity in a group such as dancing (Reddish, Fischer, & Bulbulia, 2013), exercising (Davis, Taylor, & Cohen, 2015), or praying (Fischer, et al., 2014), creates a euphoric feeling among participants, also referred to as collective effervescence (Durkheim, 1912).

Absolute majority of participants reported either using their voice or hearing others use their voice while dancing in raves. Within this group the vast majority reported using or hearing sounds comparable to CTS. The sound most participants reported to both produce and hear on the dancefloor is OOH/OOO which is the Sacral CTS. Among other, Sacral Chakra is emotionally related to creativity, sexual drive, and exploration. Physically it is related to the reproductive system, lymphatic fluids, and the pelvis. It is reasonable to assume that while dancing, which greatly involves the pelvis, this area will be strongly expressed.

Within both questionnaires people significantly reported using the EY sound (Throat Chakra) the least. While traditionally the Throat Chakra is related to self-expression, it is interesting to note that people report they express themselves as an individual less when engaged in a communal activity. However, reports relating what sounds people hear on the dance floor show similar numbers among both questionnaires between the sounds OH, AH, and EY (Solar Plexus, Heart, and Throat CTS accordingly), thus meaning there is no correlation between the sound people reported to producing the least and the sound people reported to hearing the least. Furthermore, while people reported using "other" sounds relatively often, they also reported hearing "other" sounds at the lowest likelihood on the dance floor (Figure 4A, 4B). This may be related to people's need to express themselves as individuals in an environment where they feel they are being tested, like this questionnaire and study. Another possible explanation is that some of the "other" sounds people described using are by definition less audible (for example sounds like mmm or hmmm) in comparison to shouting vowels. These sounds will be heard by fewer people on the dance floor, if at all, due to the loud music.

In response to what makes participants use their voice while dancing, responses varied mostly between three topics: 1. Music 2. Euphoric Feeling 3. Music and Euphoric Feeling. Responses relating to music tend to address bass and rhythm in particular. Low frequencies and rhythm are very present in dance and rave music which might explain this phenomenon and effect on participants. It implies that the vibratory pulsating component in the music (the bass or kick-drum) generates a mirroring inner response among participants. This mirroring effect could be related to our physicality and its dependence on its own vibratory-pulsating organ - the heart. A recent research study by Reid (2019) monitoring the effect of different types of music on in-vitro Red Blood Cells (RBC) demonstrates results showing low frequencies and rhythm have substantial effect on RBC longevity, possibly due to their similarity to heart function. Interestingly, a recent culinary experiment on cheese fermentation produced similar results relating to the positive effect of low frequencies and rhythm (Bern University of Applied Sciences, 2019). These studies and their results suggest the cellular component in living systems is excited and ignited in these sonic environments.

One key observation of this research shows the communal aspect created by self-vocalization while dancing. Evidently people are encouraged to use their voice more when they hear others do the same. Most participants report they are more than moderately encouraged to use their voice when they hear others do the same on the dance floor. These effects could be related to mirror neurons processes (Iacoboni, 2009) and support current findings showing that while visual mirror neurons effect requires both visual and target stimuli to be activated (Kilner & Lemon, 2013), in the case of an audial stimuli a sonic trigger is enough to activate mirror neurons processes (Kohler et al., 2001; Umiltà et al., 2001). Other than the physical effect of this action, this observation demonstrates the shared manifold hypothesis (Gallese, 2003) as a form of empathy, bonding, and expressing the individual with its immediate community, while creating a space that is collective and "does not segregate any subject" (p. 175).

This overall body of results of the current study is in accordance with theories and practices of both sociological and biological studies as well as spiritual leaders and traditions in vibrational medicine. Thus suggesting CTS practice is, in fact, a method by which we balance and tone our mechanical natural system as well as our emotions and subtle energy body. Founder of the Sound Healing Academy and sound healer, Nec (2017) referred to classical literature in vibrational medicine in which the vibratory mechanism activated by personal expressions such as crying, laughing, or coughing are ways to tone and balance our system (Halpern & Savary, 1985). Nec states that "When we tone we are actually toning the subtle energy body and in doing that we are allowing a release of trauma and a repatterning to take place. Vocal toning has a direct influence on the subtle energy body." Voice researcher Arenheim (2016) described the healing mechanisms that take place in our body while we use our voice as "relieving physical pain, improving mood, physical functions and general wellbeing and self-confidence." Arenheim speculated that singing has a positive effect emotionally, probably as a result of releasing specific hormones and that vibrations of tissues are known to release anti-inflammatory enzymes and help body structures renew and maintain themselves. "The vibration of the vocal folds produces acoustic energy that spreads vibrations throughout the vocal tract and spreads to tissues around it."

Another key aspect relating to CTS practice is its reported intuitive use by participants. Most people reported either producing or hearing sounds correlating to the CTS system. According to modern literature CTS as a practice seems to have developed mostly in the west, probably around the late 1970s or early 1980s, while the Bija mantras derive out of the literature in Yogic tradition. According to D'Angelo (2012), there is no eastern tradition of correlating vowels with chakras, and the choice of vowels corresponds with specific chakras developed from an intuitive general consensus. Nec (2021) notes that "In seed syllables in Sanskrit or early forms of music, such as Gregorian chanting there is a big emphasis on the vowel sounds." Nec mentions that according to Goldman, in Jewish tradition and Kabbalah the name of God '*Yahawe*' (pronounced *Yehova* or *Youhova* in Hebrew) is drawn from the vowels related to CTS. "In all these traditions the use of vowels is very prominent. There is an understanding that toning vowels can be very beneficial".

It is also interesting to note that four of the six Medicinal QiGong Healing Sounds resemble the CTS system. HAAAA (or HAWWWW) for the Heart is the most prominent. CHOOOOO for the kidneys, WHOOO for the liver, HEEEEE for the Triple Warmer sounds and their related organ functioning and occasionally positioning, highly correlate with the CTS system.

CTS form of use in different healing modalities around the world throughout the centuries, as well as different spiritual cultures, and their effect on the individual which in turn affects the group may suggest that CTS are known to all human beings intuitively. Much like a smile is a universal sign people from different cultures use intuitively, CTS may be a form of instinctive body language known to all humans and emerging from the connectivity of subtle energy fields. In its turn, intuitive perception may operate in relation to vibratory mechanisms or vibratory fields, such as Morphic Resonance, connecting our consciousness as a species

(Mcdougall, 1927 & 1930) where the individual both draws upon and contributes to its species collective memory (Sheldrake, 2012). Thus implying CTS is a form of behavioral field within morphic fields. This internal understanding of a self-organizing system is created and creates patterns of efficient energy distribution which are healing both to the individual and influencing its community.

This research is preliminary in the field and further research should be conducted in order to better investigate the findings of the current study.

Acknowledgments

This research is an independent work and was supported by the kindness of the following individuals and institutes. Special thanks and appreciation to Naomi Geffen, Vedrana, Sara Feher and the wonderful O.Z.O.R.A. family, Aramat Arenheim, Tony Nec, Maggie Langton, Sound Healing Academy, Ilana Gelernter, Ann Streichman, Maebh Murphy, Eyal Fridman and KzRadio, Female Pressure, 4DSOUND, Spatial Sound Institute, MONOM and all the people who took the time to answer the questionnaires and enrich this research with their experience. Illustrations by Giulia Palombino.

References

Anodea, J. (1987). *Wheels of life: a user's guide to the Chakra system*. Llewellyn Worldwide. Retrieved from https://www.scribd.com/book/274521485/Wheels-of-Life-A-User-s-Guide-to-the-Chakra-System

Arenheim, A. (2016). Aramat Arenheim, Retrieved from www.aramat.co.il

Bern University of Applied Sciences. Bern University of the Arts HKB. (2019). *"Cheese in Surround Sound – a culinary art experiment".* Retrieved from http://cheeseinsound.ch/wp-content/uploads/2019/03/KaeseBeschallen_mediarelease_English.pdf

Bundo, D., (2001). Social Relationship Embodied in Singing and Dancing Performances among the Baka. *African study monographs. Supplementary issue*, *26*, 85-101. doi.org/10.14989/68406

Clarke, E. DeNora, T. & Vuoskoski, J. (2015). Music, Empathy, and Cultural Understanding. *Physics of Life Reviews, 15,* 61-88. doi.org/10.1016/j.plrev.2015.09.001

Clayton, M., Sager, R., & Will, U., (2004), In Time With the Music: The Concept of Entrainment and Its Significance for Ethnomusicology. *ESEM Counter Point, 1,* 1-82. Retrieved from https://www.academia.edu/25705441/In_time_with_the_music_the_concept_of_entrainment_and_its_significance_for_ethnomusicology

D'Angelo, J. (2012). *Seed Sounds for Tuning the Chakras: Vowels, Consonants, and Syllables for Spiritual Transformation*. Inner Traditions/Bear & Company. Retrieved from https://www.scribd.com/book/351487946/Seed-Sounds-for-Tuning-the-Chakras-Vowels-Consonants-and-Syllables-for-Spiritual-Transformation

Davis, J., Taylor, & J., Cohen, E. (2015). Social Bonds and Exercise: Evidence for a Reciprocal Relationship. *PLoS ONE 10*(8): e0136705. doi.org/10.1371/journal.pone.0136705

Durkheim, E. (1912). *The Elementary Forms of Religious Life*. Dover Publications Inc. Retrieved from https://store.doverpublications.com/0486454568.html

Ferguson, J. *A Brief History of Sound Therapy*. Sivana East. Retrieved from https://blog.sivanaspirit.com/brief-history-sound-therapy

Fischer, R., Xygalatas, D., Mitkidis, P., Reddish, P., Tok, P., Konvalinka, I., & Bulbulia, J. (2014). The fire-walker's high: affect and physiological responses in an extreme collective ritual. *PloS one, 9*(2), e88355. doi.org/10.1371/journal.pone.0088355

Franka, E., Swartza, H., & Kupfer, J. (2000). Interpersonal and social rhythm therapy: managing the chaos of bipolar disorder. *Biological Psychiatry, 48*(6) 593-604. doi.org/10.1016/S0006-3223(00)00969-0

Gallese, V. (2003). The Roots of Empathy: The Shared Manifold Hypothesis and the Neural Basis of Intersubjectivity. *Psychopathology, 36*, 171–180. doi.org/10.1159/000072786

Gutgsell, K., Schluchter, M., Margevicius, S., DeGolia, P., McLaughlin, B., Harris, M., Mecklenburg, J., & Wiencek, C. (2013). Music Therapy Reduces Pain in Palliative Care Patients: A Randomized Controlled Trial. *Journal of Pain and Symptom Management, 45*(5), 822-831. doi.org/10.1016/j.jpainsymman.2012.05.008

Halpern, S., & Savary, L.M. (1985). *Sound Health: The Music and Sounds That Make Us Whole*. Harper & Row. Retrieved from https://books.google.dk/books?id=bc8YAQAAIAAJ

Harada, T., Ishizaki, F., Ito, S., Aoi, S., Miyaguchi, M., Ikeda, H., Nitta, Y., Mibu, M., Iida, T., Chikamura, C., Tamura, N., & Kohsaku, N. (2017). Study of Objective Evaluation of Effect of Psychological Relaxation Using Classic Music. *International Medical Journal, 24*(1), 31 - 33. Retrieved from https://www.researchgate.net/publication/314116067_Study_of_Objective_Evaluation_of_Effect_of_Psychological_Relaxation_Using_Classic_Music

Hausauer, N. (2008-20). *Chakra Tones*. The Energy Healing Site. Retrieved from www.the-energy-healing-site.com/chakra-tones.html

Iacoboni, M. (2009). Imitation, empathy, and mirror neurons. *Annual review of psychology, 60*, 653-70. doi.org/10.1146/annurev.psych.60.110707.163604

Jingjing, Z. (2015). *Effects of Auditory Training on Cognitive Function in Patients with Stroke*. tomatisassociation.org. Retrieved from https://tomatisassociation.org/effects-of-auditory-training-on-cognitive-function-in-patients-with-stroke

Kasprzak, C. (2011). Influence of Binaural Beats on EEG Signal. *ACTA Physica Polonica A, 119*, 986-990 doi.org/10.12693/APhysPolA.119.986

Keyes, L. (1973). *Toning: The Creative Power of the Voice*. DeVorss & Company. Retrieved from https://www.devorss.com/product/toning-the-creative-and-healing-power-of-the-voice-book-with-cd

Kilner, J. M., & Lemon, R. N. (2013). What we know currently about mirror neurons. *Current biology: CB, 23*(23), R1057–R1062. doi.org/10.1016/j.cub.2013.10.051

Kohler, E., Keysers, C., Umiltà, A., Fogassi, L., Gallese, V., & Rizzolatti, G. (2001). Hearing sounds, understanding actions: action representation in mirror neurons. *Science, 297*(5582), 846-848. doi.org/10.1126/science.1070311

Lane, J., Kasian, S., Owens, J., & Marsh, G. (1998). Binaural Auditory Beats Affect Vigilance Performance and Mood, Physiology & Behavior. *Physiology & Behavior, 63* (2), 249-252. doi.org/10.1016/S0031-9384(97)00436-8

Lee, E-J., Bhattacharya, B., Sohn, C., & Verres, R. (2012). Monochord sounds and progressive muscle relaxation reduce anxiety and improve relaxation during chemotherapy: A pilot EEG study. *Complement Therapy Med., 20*(6): 409-16. doi.org/10.1016/j.ctim.2012.07.002

Lee, J. H. (2016). The Effects of Music on Pain: A Meta-Analysis. *Journal of Music Therapy, 53*(4) 430–477. doi.org/10.1093/jmt/thw012

Mcdougall, W. (1927). An Experiment For The Testing Of The Hypothesis Of Lamarck. *British Journal of Psychology, 17*(4), 267-304. doi.org/10.1111/j.2044-8295.1927.tb00432.x

Mcdougall, W. (1930). Second report on a Lamarckian experiment. *British Journal of Psychology, 20*, 201-18. doi.org/10.1111/j.2044-8295.1930.tb00545.x

Nec, T. (2021). Sound Healing Academy. Retrieved from www.academyofsoundhealing.com

Nunner, M. (2017). *Sound Therapy and Well-Being: Some Scientific Studies.* medicinanarrativa.eu. Retrieved from https://www.medicinanarrativa.eu/sound-therapy-and-well-being-some-scientific-studies

Oster, G. (1973). Auditory Beats in the Brain, *Scientific American, 229(4), 94–102.* doi.org/10.1038/scientificamerican1073-94

Reddish, P., Fischer, R., & Bulbulia, J. (2013). Let's Dance Together: Synchrony, Shared Intentionality and Cooperation, *PLOS ONE, 8* (August 7). doi.org/10.1371/journal.pone.0071182

Redmond, L. (1997). *When the Drummers were Women*. Random House USA Inc. Retrieved from https://layneredmond.com/When_the_Drummers_Were_Women.html

Reid, J. (2010). *Rediscovering the Art & Science of Sound Healing*. Cymascope.com. Retrieved from https://www.cymascope.com/cyma_research/articles.html

Reid, J. (2019). *Testing a 2,500 year-old hypothesis: If music breathes new life into old blood cells... play on!*. Experiment. Retrieved from https://experiment.com/u/8qj2Mw

Schäfer, T., Sedlmeier, P., Städtler, C., & Huron, D. (2013). The psychological functions of music listening. *Frontiers in Psychology, 4* (Article 511).

Schamis, C. (1985). Healing processes in group dance therapy. *American Journal of Dance Therapy, 8,* pages 17–36. doi.org/10.1007/BF02251439

Sheldrake, R. (2012) *Morphic Resonance and Morphic Fields - an Introduction*. Rupert Sheldrake. Retrieved from https://www.sheldrake.org/research/morphic-resonance/introduction

Shin, H.S., & Kim, J.H. (2011). Music Therapy on Anxiety, Stress and Maternal-fetal Attachment in Pregnant Women During Transvaginal Ultrasound. *Asian Nurs Res 5*(1), 19-27. doi.org/10.1016/S1976-1317(11)60010-8

Taylor, J., Cohen, E., Davis A., & Weintraub P. (2016). *Exercising together boosts performance and forges friendships*. aeon. Retrieved from https://aeon.co/ideas/moving-in-sync-allows-the-body-to-do-things-it-couldnt-do-alone

Umiltà MA, Kohler E, Gallese V, Fogassi L, Fadiga L, Keysers C, & Rizzolatti G. (2001). I know what you are doing. a neurophysiological study. *Neuron, 31*(1), 155-65. doi.org/10.1016/s0896-6273(01)00337-3

Uvnäs-Moberg, K., & Petersson, M. (2005). Oxytocin, a Mediator of Anti-stress, Well-being, Social Interaction, Growth and Healing. *Z Psychosom Med Psychother, 51*(1), 57-80. doi.org/10.13109/zptm.2005.51.1.57

Von Muggenthaler, E. (2001). The felid purr: A healing mechanism? *The Journal of the Acoustical Society of America, 110*(5). doi.org/10.1121/1.4777098

Wells, A., Hakanen, E. (1991). The Emotional Use of Popular Music by Adolescents. *Journalism Quarterly, 68*(3), 445-454. doi.org/10.1177/107769909106800315

Author's Note

None of the above should deflect from the importance of protecting one's valuable and precious ears. Necessary precautions to protect your hearing are always important and highly recommended when listening to loud music in raves, festivals and parties. One of the best self-care investments one can make is a pair of high-quality custom made ear plugs which monitor high volumes without affecting the frequency range and your enjoyment of the music. On a more personal note, I'd like to add a shout out to my precious community. Raves of electronic music, tribal gatherings, festivals and popular music concerts and parties are often regarded as an inferior form of entertainment, orchestration, and musicality. The overall effect and significance of these events is usually dismissed in comparison to use of softer, more "harmonic" instrumentation. This perception deprives people of a valuable tool for self-expressions and self-healing - "Last night a DJ saved my life" is more than just a catch phrase. DJs, if the party is loud, you are doing it right! Ravers, have no shame expressing yourselves on the dance floor, it is a beautiful act of self-healing and ommunal communication. Our culture is beautiful and meaningful. The joy we experience when we gather and listen to the music we love resonates from our most inner provenance outwards to uplift our spirit, heal our soul, balance our physicality and communicate with our fellow beings on the dance floor. We are one united in sound. Rave On!

Factors Affecting the Mental Health of Adult Musicians with Learning Disabilities: Conclusions from a Qualitative Study

Sheerin Hosseini

Abstract

This paper aims to describe the results of a qualitative dissertation study conducted by Hosseini (2020) in which the lived experiences of ten successful adult musicians with dyscalculia – a mathematical learning disability - were explored. Numerous themes emerged in this study, including (1) Self-Esteem, (2) the Impact of Teachers, (3) Physical and Mental Health, (4) Math Anxiety, (5) Math Avoidance, (6) Perseverance, and (7) Empathy.

Participants described experiencing physical and mental health issues as a result of early educational experiences with unsupportive teachers in academic and musical contexts, which led to feelings of anxiety, stress, and inadequacy. Understanding these connections is vital for medical professionals, parents, and teachers to create conducive environments for individuals with learning disabilities to feel supported, valued, and receive the help they need to succeed.

Keywords: Music Learning, musicians, dyscalculia, mathematical learning disabilities, mental health

Introduction

Adamek and Darrow (2018) defined a learning disability as "a neurological condition that affects a student's ability to store, process, or produce information" (p. 228). Two common learning disabilities affecting approximately 5%-8% of the population include dyslexia and dyscalculia. According to the *Diagnostic Statistical Manual of Mental Disorders* (5th ed.; *DSM-5*; American Psychiatric Association, 2013), dyslexia is "characterized by problems with accurate or fluent word recognition, poor decoding, and poor spelling abilities" (p. 67). Dyscalculia, on the other hand, is "characterized by problems processing numerical information, learning arithmetic facts, and performing accurate or fluent calculations" (p. 67).

Research conducted in music education and music therapy indicates positive social, emotional, and cognitive advantages for individuals with learning disabilities. For example, Adamek and Darrow (2018) cited several studies illustrating how music can be helpful in the classroom for children with learning disabilities. Some benefits of music participation for students with special needs included allowing students opportunities to experience success (Martin, 1983); learning compensatory skills (Cahn, 1987); expressing oneself in different ways (McCarthy, 1985); making friends (Brodeur, 1989); and to use mnemonic devices and adopt new ways of learning (Claussen & Thaut, 1997; Gfeller, 1986; Wolfe & Hom, 1993).

Furthermore, listening and performing music has been shown for a wide range of medical conditions and disabilities. Dingle et al. (2021) additionally examined how participating in music activities affected health and well-being. Their scoping review of studies exploring psychosocial mechanisms identified 63 studies indicating how listening to music, performing music, or engaging in movement and dance activities led to improvements among participants of diverse ages and backgrounds in physical, emotional, or mental health in various medical and educational settings. One specific study reviewed by Dingle et al. was conducted by Wilson et al. (2019). They identified 37 adults with mild to profound learning disabilities with little access to music activities. The researchers found that participants experienced increased enjoyment and enthusiasm when engaging with music despite little prior exposure to music. The participants also experienced increased self-confidence, happiness, relaxation, and comfort in unfamiliar environments and social situations.

However, while engaging with music provides fulfillment and enjoyment for many in a wide range of conditions, learning to perform music may also prove difficult for individuals with or without learning disabilities because learning disabilities are domain-specific (American Psychiatric Association, 2013). For individuals considered to be "twice-exceptional," this may especially be true. This phenomenon occurs where a learning disability is present in one academic area, and talent may be prevalent in another domain such as music (Hammel & Hourigan, 2017; Rief & Stern, 2010; Brody & Mills, 1997). Therefore, some aspects of learning to perform music may come relatively easy for these unevenly gifted individuals, while other musical concepts are more complicated to understand.

As a result of this discrepancy, some individuals with learning disabilities may experience stress and anxiety. For example, addressing a specific learning disability, dyslexia, Scott (2004) wrote ... "stress and anxiety are direct and indirect responses to actual and anticipated failure, fear, frustration, suppressed anger, and the erosion of confidence" (p. 157). For children, teens, and adults with learning disabilities, Rief & Stern (2010) listed low self-esteem, low tolerance for frustration, stress, anxiety, and acting out as occurring side-effects for some individuals.

Scant research has been conducted on the lives of adult musicians with learning disabilities regarding their musical challenges and strengths and how learning and performing music affects their mental and emotional health. Nelson (2014) explored the music coping strategies of five musicians with dyslexia to gain insight into the perceptions of professional musicians as they reflected on their music learning experiences throughout their lives. Two research questions were posed: (1) What strategies have five professional musicians used to overcome the challenges associated with dyslexia? and (2) What are the perceived abilities and traits the participants believe they have developed in music because of their diagnoses of dyslexia?

Nelson (2014) found that the participants utilized various strategies, including multisensory learning and teaching, small group and private instruction, technology such as recordings, isolating musical components, and learning and performing other music genres with less emphasis on standard music notation. In terms of the participants' strengths and weaknesses, four out of the five participants felt their dyslexia contributed to their musical strengths. The common weaknesses expressed by the participants included low self-esteem and music reading ability.

Using Nelson's research as a foundation, this current study (Hosseini, 2020) examined how a specific learning disability, dyscalculia, affected ten adult musicians' daily lives, education, and careers. The research questions were (1) How did the participants' dyscalculia affect their learning experiences in school, daily life, and career? (2) How did the participants' dyscalculia affect their music learning experiences in school and their professional lives? and (3) Do musicians with dyscalculia develop or use specific compensatory strategies to overcome learning challenges or utilize their musical strengths in unique ways?

Numerous themes emerged in this study, including (1) Self-Esteem, (2) Impact of Teachers, (3) Physical and Mental Health, (4) Math Anxiety, (5) Math Avoidance, (6) Perseverance, and (7) Empathy. One purpose of this research study was to discover how the participants' musical experiences impacted their physical and mental health in relation to the other six themes. Understanding these connections is vital for medical professionals, parents, and teachers to create conducive environments for individuals with learning disabilities to feel supported, valued, and receive the help they need to be successful throughout their life.

Permission to conduct this study was granted by the Institutional Review Board (IRB) at the University of Miami. Data was gathered through interviews using video conferencing software such as Facetime, Skype, and Facebook Messenger. The interviews were transcribed verbatim, and the participants member-checked the write-ups. The participants were all involved in music-related fields including, singing, songwriting, music education, musicology, and instrumental performance. A brief description of the participants is included in Table 1. Pseudonyms were used to maintain confidentiality.

Table 1
Description of Participants

Skylar	A college senior majoring in classical percussion performance with mathematical difficulties from the Pacific Southwestern part of the United States
James	A general music teacher from the Southeastern part of the United States with ADHD, dyslexia, and mathematical difficulties
Emily	A general music teacher from the Mid-Atlantic/Northeast region of the United States with ADHD and mathematical difficulties
Hannah	A general music teacher from the Southeastern part of the United States with mathematical difficulties
Paul	A music education professor and musicologist from the Southeastern part of the United States with mathematical difficulties
Kristin	A musicologist from the Southeastern part United States with mathematical difficulties
Samantha	A music educator, vocal coach, and graduate student from the Midwest with mathematical disabilities
Lisa	A singer-songwriter who lives in the Southeastern part of the United States with dyslexia and dyscalculia
Kelly	A singer-songwriter from the Southeastern part of the United States with ADHD and dyscalculia
Jared	A music educator and rock musician from the United Kingdom with dyslexia and mathematical learning difficulties

Post-Secondary Outcomes of Adults with Learning Disabilities

Several researchers have expressed interest in the post-secondary outcomes of individuals affected by learning disabilities, while others focused on dyslexia. In examining the literature pertaining to this topic, there were several underlying trends that will be weaved in this article: (1) individuals with learning disabilities may experience a sense of inadequacy in comparison to their peers (Ebo, 2016; Heiman & Kariv, 2004; Heiman & Precel, 2003; Masover, 1999; McNulty, 2000; Stage & Milne, 1996; Vogel et al., 2007), (2) obtaining a college education is considered a significant accomplishment for individuals with learning disabilities (Madaus, 2006; Madaus et al., 2001), (3) individuals with learning disabilities feel that it is through a significant amount of effort and the belief in their ability to achieve contributes to their overall success both academically and in the workplace (Ebo, 2016; Greenbaum et al., 1995; Heiman & Kariv, 2004; Heiman & Precel, 2003; Madaus, 2006; Madaus et al., 2001; Reis et al., 2000), and (4) successful individuals with learning disabilities develop coping strategies to overcome challenges (Greenbaum et al., 1995; Heiman & Kariv, 2004; Heiman & Precel, 2003; Reis et al., 2000; Stage & Milne, 1996), These four trends were also apparent in the participants' lives in this study and will be used as headings to discuss to relevant themes.

Self-Esteem

The first trend identified in this study was that individuals with learning disabilities felt a greater sense of inadequacy compared to their peers without learning disabilities, which impacts their self-esteem (Ebo, 2016; Heiman & Kariv, 2004; Heiman & Precel, 2003; Masover, 1999; McNulty, 2000; Stage & Milne, 1996; Vogel et al., 2007). Stets and Burke (2014) cited Rosenberg et al.'s (1995) definition of self-esteem. They defined self-esteem as "the negative or positive attitude individuals have of themselves" (p. 409). Low feelings of self-esteem may result due to music learning challenges.

For example, research has indicated that individuals with learning disabilities may struggle to understand some mathematically related musical concepts, making them feel inferior to their peers (Ganschow, Lloyd-Jones, and Miles, 1994; Nelson, 2014). In this study, many participants experienced difficulties with musical tasks such as sight-reading, rhythm, coordination, and music theory. For example, Lisa stated that although she could "play guitar, make chords and write songs," she felt that she does not play the guitar with as much facility as she would like due to issues with hand-eye coordination. She declared. "I don't really think of myself as a musician like I think of my friends as musicians" because "they play instruments." She continued:

> Whenever I'm meeting someone, I'm like yeah, I'm a musician too, well, not really a musician because I don't play easily. I sort of tap out with the guitar. I can practice, practice, practice, but it's just always been a love-hate relationship for me personally playing.

Jared, a percussion teacher, sometimes contemplated whether or not teaching was the right profession since he sometimes struggled with understanding rhythmic concepts. He revealed:

> Things sometimes will get frustrating, and sometimes I feel like you question yourself whether you're in the right line of work and should be doing something else. But then, based on my dyslexia and things, well, can I do those other things? Do I have the skills and abilities to be able to learn those being the age I am?

In addition to difficulties with certain musical tasks, the participants' interactions with their teachers in school also contributed to how the participants felt about themselves.

Impact of Teachers

Another finding of this study was that positive and negative interactions with teachers impacted the participant's self-esteem. For example, empathetic teachers made a positive difference in the participants' lives, while unsupportive teachers had a detrimental effect on the participant's self-esteem. Humphrey (2003) stated that in research interviews with children with dyslexia, it was clear that "the influence of teachers as significant others was an important factor in the low-self-esteem exhibited by so many of them" (p. 131). For instance, James described his challenges with keeping a steady beat, counting rhythm, getting lost in the music, and sight-reading. He mentioned specific challenges related to these tasks. He stated:

> When I was in college, I was fussed at because I couldn't keep a steady beat at times because I went with what I felt. In fact, it was so bad that one of my professors said I needed to change my major, and I'd never be a music teacher.

Other participants shared experiences where their music teachers were emotionally supportive and helped them overcome music learning challenges, leading to an increased sense of self-esteem regarding their musical ability. Samantha shared her experiences with her teachers:

> I had an excellent middle and high school choir director from sixth through tenth grade. He was the same person, and then when he departed, ...he was also our church choir director, and so I took his position. And that's when I started voice lessons. That's when I started acting lessons, and that teacher was my rock. She was my emotional rock all the way through high school, and she's the reason I made it through being in a competitive show choir and auditioning for college, and all of that.

Finally, James shared his experience with a teacher who made an effort to work with him one-on-one participating in the marching band and his struggle to the tell his left from his right.

> My band director would also spend time working with me, before or after class, before or after rehearsal. So, he gets out there and says, "Do this. Do this. Let's try this." At one point, what we ended up doing, and I still do this when I hit recorders and stuff. We ended up putting a band-aid on my left hand because the band-aid, I could feel, and that would help me remember left and right.

These examples demonstrate that music teachers play an essential role in the mental and emotional well-being and success of students with learning music.

Mental and Physical Health

Shessel and Reiff (2011) found that other reported difficulties of individuals with learning disabilities included daily living, experiencing the impostor phenomenon, social isolation, and emotional health damage. Monitoring emotional health problems among children with and without learning disabilities is profoundly important. Huntington and Bender (1993), stated "recent evidence has suggested that adolescents with learning disabilities demonstrate higher anxiety levels, more frequent and more serious bouts of depression, and higher rates of suicide than adolescents without disabilities" (p.159).

Due to this fact, it was not surprising that some participants in this study experienced psychological or physical health challenges. Hannah and Samantha, for example, discussed how their rigorous university departments were designed to weed out students who did not possess the necessary qualities to succeed musically in the program. Due to the demanding nature of the universities she attended, Samantha affirmed, "if I didn't fill in those holes, I was done because they separate the wheat from the chaff." She continued to state, "It's like with every college student who is horribly over-committed and is not getting enough sleep." She noted, "I also had a physical meltdown. They were concerned that I would have heart issues because I was so overtired."

Stress experienced by individuals with learning disabilities may be known as achievement stress. Achievement stress tends to go undetected but generally worsens as children advance through school (Hill and Wigfield, 1984). Rubenzer (2021) stated that "achievement stress may result in academic failure, behavioral or emotional problems, drug abuse, health problems, and even suicide."

Math Anxiety

In addition to feelings of low self-esteem, the participants relayed experiences with different types of anxiety, most notably math anxiety and test-taking anxiety. Based on a thorough overview of research conducted on math anxiety, Dowker et al. (2019) asserted that "there is a strong bidirectional relationship between math anxiety and low attainment in mathematics." Anxiety associated with math tests was also a concerning topic among many participants. Hannah, for example, shared her experience with test-taking:

> I had panic attacks. I don't have them as much anymore, but a throat closing off sensation...hot, kind of feeling flushed. That's usually how I feel when taking a math test. I can remember taking one in particular where the air conditioner sounded so loud that I couldn't focus, and it really wasn't that loud. It was just I couldn't focus. I was so worked up about the test that that's all I could think about was the humming of the air conditioner and throat closing off, being really upset kind of deal.

Clearly, various forms of anxiety affected the participants in stressful situations, such as taking exams and performing in front of others. Stress was also a factor for high achieving participants when they encountered challenges in school or the workplace. Many participants compensated for their anxiety around numbers by avoiding math altogether. This theme will be discussed in the next section.

Math Avoidance

Choe et al. (2019) found that one consequence of mathematics anxiety is mathematics avoidance. These authors posited that math avoidance behaviors include taking fewer classes related to science, technology, engineering, and mathematics courses than people who do not experience math anxiety. Math avoidance, another theme, was evident in the participants' comments regarding their daily lives and careers. Interestingly, several participants relied on their significant others to handle everyday matters related to money. Jared noted:

> I was eating a meal, me and my girlfriend. We finished up, and we paid the tab, and you know, she did the math straight away (snaps). It's like, okay, I'll never compete with that because she might as well do it because she's so much quicker at that than I am.

Many other participants revealed that they rely on their significant others to help them with mathematical related tasks. For instance, Emily, a general music teacher, remarked that since her husband has a degree in physics, "he does any and all the math necessary." As a result, the only math-related task she needs to perform is grading. What became evident throughout this study was that in many ways, the participants' life challenges resulted in an increased sense of empathy for others.

Empathy

Several participants in this study cited empathy as a strength. This discovery was consistent with the findings of Shessel and Reif's (2011) study in that the participants felt increased sensitivity to others and developed the desire to help others. Kelly stated:

> The helpful thing. A strength of mine is I tend to see where people are coming from. I won't get into politics or anything like that, but I can look at different people who have these very strong opinions that are very different. And I don't feel very judgmental about it because I feel like everybody is coming from a different place.

Furthermore, the participants' learning challenges as children and young adults, particularly among the five music educators in this study, led to the ability to sympathize with their students who struggled to understand certain concepts. Paul explained that his office door is open to students most of the time if they're working on homework or are having trouble stating, "my policy is I'll help you do your homework. I encourage them to see me if they don't understand something." Paul acknowledged that he wants to help his students in this way because he knows his brain works differently. As a result, these educators understood the importance of fostering learning environments where struggling students felt supported rather than left out. For example, Jared noted:

> Well, as a teacher, you know some people, some kids need some encouragement in the right way. And then some people, you know, they need a stern word of like you need to pull up your bootstraps and get to work here. And then you know, some people need an arm put around them and just saying you know, you did really well. And especially being dyslexic, and being a teacher, I can sort of read that now a bit more, and I can spot that when I'm teaching someone, that I can see they're struggling, and you know they're climbing Mount Everest here in trying to do this. And you know they need a real pat on the back and say, well done.

Perseverance

Perseverance emerged as a key element responsible for the participants' success as musicians and teachers. For example, the second trend described in several studies indicated that individuals with learning disabilities are largely successful in academic and workplace settings despite any protentional challenges due to effort and hard work (Ebo, 2016; Greenbaum et al., 1995; Heiman & Kariv, 2004; Heiman & Precel, 2003; Madaus, 2006; Madaus et al., 2001; Reis et al., 2000). In this current study, many participants spoke about their resiliency in the face of adversity. James provided one example in which he worked diligently to improve his trombone playing skills. He explained, "At the beginning of that freshman year, I was labeled probably the weakest player in the high school band." As a result of his hard work, James went from being the weakest player his freshman year to the strongest player his junior year without private lessons. James even won an award two years in a row for the most improved player. He shared:

> So, that gave me a lot of confidence, and that's voted on by the members of the band, is the way that is. The director's out of that one totally. So, I won it two years in a row. So, that tells you that I went from that freshman year, where I was the worst player, into my junior year, I was the best player in the band.

Lisa also reflected on her determined mindset, which allowed her to overcome insecurities surrounding her musical challenges and keep pursuing her musical goals. She remarked:

> So I think just this kind of intuition or instinct— just keep at it. Lisa, just keep at it. This is what you're hearing in your head. Don't let someone tell you it's this because a lot of people are trying to help you work something out. So I guess that coping thing was kind of stubbornness.

The third trend that corroborated with the findings of others regarding the post-secondary outcomes was that obtaining a college education represented a significant accomplishment for students with learning disabilities, resulting in their ultimate success (Greenbaum et al., 1995; Madaus et al., 2001, & Madaus, 2006). In this current study, most participants obtained a college education and pursued careers that allowed them to reach their personal goals. Out of ten participants, four earned a doctoral degree, and four obtained bachelor's degrees.

Several participants in this study noted that they worked very hard to overcome musical challenges to do well in their college music courses. Samantha noted that she struggled to understand syncopation and sometimes made rhythmic mistakes when playing the piano, singing, or sight-reading. She explained, "Everything would be fine," "... and then all of a sudden, it would be as if my brain would flip out and stop being able to decode the notation. And again, it would be that stopping sensation." Samantha overcame this challenge by "sitting for 3 hours a day, six days a week in a practice room practicing at 1/4 tempo" until she completely understood it.

Hannah also described instances in which she lacked the basic theory skills to excel in her music theory classes because she felt the music program at her high school did not prepare her well. For example, Hannah revealed that understanding music theory was also challenging because "it is kind of mathematical too," and said, "that kind of bothers me." Despite her struggles, out of twenty-two vocalists in the class, only three finished, and she was among them.

In research studies exploring the experiences of college students with learning disabilities, it was found that elements such as motivation, dedication, the development of excellent work habits, as well as the belief in their ability to succeed were essential to their academic success (Ebo, 2016; Greenbaum et al., 1995; Heiman & Kariv, 2004; Heiman & Precel, 2003; Madaus, 2006; Madaus et al., 2001; Reis et al., 2000). Furthermore, successful students expended more effort than their peers (Reis et al., 2000; Stage & Milne, 1996).

On the other hand, in many cases, researchers indicated that the extreme effort required to overcome challenges caused by learning disabilities can cause emotional and mental health issues. Concerning academic and workplace stressors, Masover (1999) found that some of the concerns of individuals with learning disabilities included low self-esteem, difficulties maintaining employment, being highly emotional, and experiencing intense anxiety after taking exams. Participants with learning disabilities, however, have shown resilience by developing coping strategies to deal with both academic and musical challenges.

The fourth and final trend identified in research on the post-secondary outcomes of individuals with learning disabilities was that successful individuals with learning disabilities develop coping strategies to overcome challenges (Greenbaum et al., 1995; Heiman & Kariv, 2004; Heiman & Precel, 2003; Reis et al., 2000; Stage & Milne, 1996). Given that some musicians with learning disabilities struggle to understand mathematical and visual-spatial-related musical concepts (Ganschow, Lloyd-Jones, and Miles, 1994; Nelson, 2014), it was interesting how the participants developed music coping strategies and utilized their musical strengths in unique ways. These coping strategies pertained to sight-reading, playing the piano, rhythm, counting measures rest, music theory. Left-right confusion may sometimes affect motor skills, marching, or performing choreography (Blair & McCord, 2017). The participants used a variety of strategies, including listening to a recording, memorizing music, slow practice, repetition, multisensory learning strategies, finger counting systems, and utilizing rhythm syllables instead of numbers to count rhythm. Samantha offered an example of her process for deciphering complex rhythms.

> I would subdivide, and I would literally tap on my body the sixteenth notes until I could feel it, and then I would march it, and I would choreograph it, and I would conduct it. And I would speak it, and I would sing it, and I played it on the piano. So every method of input that I could, I was using in order to learn those things. And now they are much, much faster.

This research study revealed that musicians with learning disabilities might struggle with musical concepts like sight-reading, counting rhythm and measures rest, and music theory. Other difficulties may impact their ability to play an instrument or engage in movement activities due to coordination difficulties and left-right confusion. However, musicians with learning challenges can overcome these barriers through hard work, effort, belief in their ability to succeed, develop coping strategies, and utilizing their musical strengths. However, it is possible that the stress of living with a learning disability and the associated challenges that may exist result in these individuals becoming depressed, anxious, and lacking self-confidence, which is why it is essential to remain mindful of this potentiality.

Conclusions

This study explored the experiences of adult musicians with dyscalculia as it pertained to their physical and mental health, and how these experiences impacted the participants' self-esteem in various and numerous ways. These findings corroborated with the conclusions of other researchers discussed in the research literature on the post-secondary outcomes of individuals with learning disabilities. The DSM-5 (American Psychiatric Association, 2013) stated that some functional consequences of learning disabilities included "high levels of psychological distress and poorer overall mental health" (p. 73). Aro et al. (2019) found differences between 430 adults with diverse subtypes of learning disabilities compared with 2,1249 adults without learning disabilities on adult-age mental health, education, and employment with learning disabilities and control groups in all outcomes. Therefore, it is crucial to consider that individuals with learning disabilities may be more prone to mental health issues.

Understanding how living with learning disabilities may affect an individuals' mental and emotional well-being is crucial. Unlike most participants in this study who were able to develop coping strategies to overcome musical, educational, and career challenges, other individuals with learning disabilities may not reach that level of success. Obtaining a college education provides more opportunities upon graduation for individuals with learning disabilities than those who did not attend college. Masover (1999), for instance, found that the participants' levels of education directly affected their career choices. All of the respondents in Masover's study, except one, could not meet their career goals. According to Madaus (2006), a successful post-secondary college experience can make a crucial difference in workplace settings.

The findings revealed in this current study indicated that early educational experiences with unsupportive teachers in academic and musical contexts led to the participants experiencing feelings of anxiety, stress, and inadequacy which lasted throughout their lives. In the case of the participants, a key factor to their success was due to a combination of their determination not to give up, a belief in their ability to succeed, and the ability to develop coping strategies to overcome challenges. For this reason, educators and other adults must foster conducive learning environments where children can gain the skills they need to boost their self-confidence, advocate for themselves, and seek appropriate accommodations when necessary. When music teachers understand, empathize, and advocate for children with learning disabilities, lifelong enjoyment of music is possible as numerous participants in this study declared their love for music despite any challenges they experienced. James, for instance, stated, "I just fell in love with it." Ultimately, their passion for music fueled many of the participants' desire to reach their musical and personal goals.

From a more general standpoint, parents and health professionals must also acknowledge how learning disabilities can impact a person's physical and mental well-being. This knowledge will ensure that children and young adults are supported, leading to less negative emotions and more successful, productive lives into adulthood. A key takeaway from the research discussed in this research article is that it is vital for parents, educators, and medical professionals to identify potential negative contexts that may trigger mental health issues in individuals with learning disabilities. Most importantly, it is important to understand the effect that a supportive home and learning environment can have on an individual's overall emotional, social, and mental well-being with learning disabilities. Further research is needed to explore how learning disabilities impact the mental health of beginning, intermediate, and professional musicians.

References

Adamek., M., & Darrow, A.-A. (Eds.). (2018). *Music in special education* (3rd ed.). American Music Therapy Association.

American Psychiatric Association. (2013). *Diagnostic and statistical manual of mental disorders* (5th ed.). https://doi.org/10.1176/appi.books.9780890425596

Aro, T., Eklund, K., Eloranta, A. K., Närhi, V., Korhonen, E., & Ahonen, T. (2019). Associations between childhood learning disabilities and adult-age mental health problems, lack of education, and unemployment. *Journal of Learning Disabilities*, *52*(1), 71–83. https://doi.org/10.1177/0022219418775118

Blair, D. V., & McCord, K. A. (2016). *Exceptional music pedagogy for children with exceptionalities: International Perspectives*. Oxford University Press. https://doi.org/10.1093/acprof:oso/9780190234560.001.0001

Brodeur, J. (1989). Music as a therapeutic tool to increase social skills in the learning disabled children. *Canadian Music Education*, *31*(2), 88-107.

Brody, L. E., & Mills, C. J. (1997). Gifted children with learning disabilities: A review of the issues. *Journal of Learning Disabilities*, 30(3), 282-286. https://mycty.jhu.edu/mycty/docs/giftedldissues.pdf

Cahn, L.D. (1987). *Teaching piano students with learning disabilities.* (ERIC Document Reproduction Service No. ED286298)

Choe, K. W., Jenifer, J. B., Rozek, C. S., Berman, M. G., & Beilock, S. L. (2019). Calculated avoidance: Math anxiety predicts math avoidance in effort-based decision making. *Science Advances*, *5*(11), eaay1062. https://doi.org/10.1126/sciadv.aay1062

Claussen, D., & Thaut, M. (1997). Music as a mnemonic deice for children with learning disabilities. *Canadian Journal of Music Therapy*, *5*(1), 55-66.

Dingle, G.A., Sharman, L. S., Bauer, Z., Beckman, E., Broughton, M., Bunzli, E., Davidson, R., Draper, G., Fairley, S., Farrell, C., Flynn, L. M., Gomersall, S., Hong, M., Larwood, J., Lee, C., Lee, J., Nitschinsk, L., Peluso N., Reedman, S. E., …and Wright O. R. (2021, September 8). How do music activities affect health and well-being? A scoping review of studies examining psychosocial mechanisms. *Frontiers in Psychology.* doi: 10.3389/fpsyg.2021.713818

Dowker, A., Cheriton, O., Horton, R., & Mark, W. (2019). Relationships between attitudes and performance in young children's mathematics. *Educational Studies in Mathematics, 100*(3), 211–230. https://doi.org/10.1007/s10649-019-9880-5

Ebo, B. (2016). *Understanding the experiences of college students with learning disabilities* (Publication No. 10168267) [Doctoral dissertation, Northeastern University]. ProQuest Dissertations and Theses Global.

Ganschow, L., Lloyd-Jones, J., Miles, T. R. (1994). Dyslexia and musical notation. *Annals of Dyslexia*, 44, 185-202.

Gfeller, K. E. (1986). Music mnemonics for learning disabled children. *Teaching Exceptional Children*, *19*(1), 28-30.

Greenbaum, B., Graham, S., & Scales, W. (1995). Adults with learning disabilities: Educational and social experiences during college. *Exceptional Children*, *61*(5), 460–471. https://doi.org/10.1177/001440299506100505

Hammel, A. M., & Hourigan, R. M. (2017). *Teaching music to students with special needs: A label free approach* (2nd ed.). Oxford University Press.

Heiman, T., & Kariv, D. (2004). Manifestations of learning disabilities in university students: Implications for coping and adjustment. *Education*, *125*(2), 313–324.

Heiman, T., & Precel, K. (2003). Students with learning disabilities in higher education. *Journal of Learning Disabilities*, *36*(3), 248–258. https://doi.org/10.1177/002221940303600304

Hill, K. T., & Wigfield, A. (1984). Test anxiety: A major educational problem and what can be done about it. *The Elementary School Journal, 85*(1), 105–126. HYPERLINK "https://doi.org/10.1086/461395"https://doi.org/10.1086/461395

Hosseini, S. (2020). *The lived experiences of adult musicians with dyscalculia: A heuristic inquiry.* (Doctoral Dissertation) https://scholarship.miami.edu/discovery/fulldisplay/alma991031524480802976/01UOML_INST:ResearchR repository

Huntington, D. D., & Bender W.N. (1993). Adolescents with learning disabilities at risk? Emotional well-being, depression, suicide. *Journal of Learning Disabilities*, 26(3), 159-166.

Humphery, N. (2003). Facilitating a positive sense of self in pupils with dyslexia: The role of teachers and peers. *Support for Learning*, *18*(3), 130-136. doi:10.1111/1467-9604.00295

Madaus, J. (2006). Employment outcomes of university graduates with learning disabilities. *Learning Disability Quarterly*, *29*(1), 19–31. https://doi.org/10.2307/30035529

Madaus, J. W., Foley, T. E., McGuire, J. M, & Ruban, L. M. (2001). A follow-up investigation of university graduates with learning disabilities. *Career Development and Transition for Exceptional Individuals*, *24*(2), 133–146. https://doi.org/10.1177/088572880102400204

Martin, M. (1983). Success teaching spelling with music. *Academic Therapy*, *18*(4), 505-507.

Masover, E. J. (1999). *Adult outcomes of living with dyslexia* (Publication No. 1395025) [Doctoral dissertation, Villanova University]. ProQuest Dissertations and Theses Global.

McCarthy, W. G. (1985). Promoting language development through music. *Academic Therapy*, *21*(2), 237-242.

McNulty, M. A. (2000) *Dyslexia and the life course* (Publication No. 9979953) [Doctoral dissertation, Institute for Clinical Social Work]. ProQuest Dissertations and Theses Global.

Nelson, K. P. (2014). *Successful strategi*es of *individuals with dyslexia in the field of music: A comparative case study*. (Publication No. 3581077) [Doctoral dissertation, Boston University]. ProQuest Dissertations and Theses Global.

Rief, S. F., & Stern, J. M. (2010). *The dyslexia checklist: A practical reference for parents and teachers.* San Francisco, CA: Jossey-Bass.

Reis, S. M., McGuire, J. M., & Neu, T. W. (2000). Compensation strategies used by high ability students with learning disabilities. *Gifted Child Quarterly*, *44*(2), 123–134. https://doi.org/10.1177/001698620004400205

Rosenberg, M., Schooler, C., Shoenbach, C., and Rosenberg F. (1995). Global self-esteem and specific self-esteem: Different concepts, different outcomes. *American Sociological Review*, *60*(1), 141-156.

Rubenzer, R. L. (2021). Stress management for the learning disabled. *LD Online: All About Learning Disabilities and ADHD*. Retrieved from http://www.ldonline.org/article/6153.
Scott, R. (2004). *Dyslexia and counseling.* Whurr, London.

Shessel, I., & Reiff, H. B. (1999). Experiences of adults with learning disabilities: Positive and negative impacts and outcomes. *Learning Disability Quarterly, 22*, 305-316.

Stage, F. K., & Milne, N. V. (1996). Invisible scholars: Students with learning disabilities. *The Journal of Higher Education*, *67*(4), 426–445. https://doi.org/10.1080/00221546.1996.11780268

Stets, J. E., & Burke, P. J. (2014). Self-esteem and identities. *Sociological Perspectives*, *57*(4), 409–433. https://doi.org/10.1177/0731121414536141

Vogel, S. A., Murray, C., Wren, C., & Adelman, P. R. (2007). An exploratory analysis of employment-related experiences of educators with learning disabilities. *Educational Considerations*, *34*(2), 15–20. https://doi.org/10.4148/0146-9282.1206

Wilson, G. B., & MacDonald, R. A. R. (2019, June 28) The social impact of musical engagement of young adult with learning difficulties: A qualitative study. *Frontiers in Psychology.* https://doi.org/10.3389/fpsyg.2019.01300

Wolfe, D., & Hom, C. (1993). Use of melodies as structural prompts for learning and retention of sequential verbal information by preschool students. *Journal of Music Therapy*, *30*(2), 100-118.

Healing With Sound and Vibration With Tuning Forks

June Leslie Wieder

Abstract

Music has been used for healing. Not only musical instruments and singing, but also humming, chanting, bowls, bells, gongs, and tuning forks. Many sound healers use tuning forks. I have studied the use of the vibration of tuning forks to help the body get back in tune. Is it possible, I thought, that some kind of energy echoes between the primary and secondary curves of the spine in order to maintain the structural and neural integrity of the spine and nervous system? This led to a long period of research into the resonances of the spine. Using tuning forks, I tested the frequencies of each vertebra of the spine and found that each vertebra responded to specific tones. This research is described in my book *Song of the Spine*. The spine is similar to a piano that needs to be tuned. Using the vibration of tuning forks is not practical for extended periods of time because tuning forks need to be struck repeatedly. So, I commissioned the development of an electronic tuning fork. The first was the Song of the Spine BoneToner. Next came the Harmonic Resonator for the entire body. And finally, the Adjustable Electronic Tuning Fork (ETF). I am continuing my research with the Adjustable ETF. The vibrational treatment can be used for different neuromuscular conditions such as depression, chronic and acute pain, multiple sclerosis (MS), and autism.

Keywords: Electronic Tuning Fork, tuning forks, resonance vibration, vibrational healing, harmonic resonance

Introduction

Throughout history music and vibration have been used for healing. Some healers use musical instruments, others use bowls, bells, gongs, and other resonant "sound-makers." Some practitioners use their voices for singing, some for humming, and others use their voice for chanting or toning. For many years in my practice, I have used tuning forks as a vibrational and sound-medicine tool. I studied the effects of the vibration of tuning forks on the spine with the goal of bringing the body back in balance. Is it possible, I thought, that some kind of energy echoes between the primary and secondary curves of the spine to maintain the structural and neural integrity of the spine and nervous system? The spine has two curves. We are born with the primary, or kyphotic, curve. The secondary curve, called the lordotic curve, develops as we sit and learn to walk. The primary curve, which consists of the thoracic or dorsal vertebrae, has a total of 12 vertebrae. The secondary curve, the lumbar and cervical vertebrae, also has 12 vertebrae.

A musical octave contains 12 semi-tones. I became very curious to know if the 12 vertebrae of each spinal curve corresponded to the 12 semi-tones of an octave. This led to a long period of research into the resonance frequencies of the spine. More than three thousand individuals were tested using tuning forks and muscle testing.

As a musician I was familiar with tuning forks, and as a chiropractor I was familiar with muscle testing. During my training as a chiropractor, I became familiar with the diagnostic system called applied kinesiology. This system uses muscles testing as a primary feedback mechanism to evaluate how the body reacts to stimuli applied to the nervous system. Applied kinesiology has been defined as functional neurology, the study of how the nervous system controls muscle and body functions. Initially this test was used for polio patients to determine if muscles were strong or weak.

What is actually tested is the nervous system that controls muscle function. Dr. George J. Goodheart Jr., a chiropractor, discovered that this testing can be used as a general diagnostic tool to determine neuromuscular dysfunction.

Muscle testing is both an art and science. In order to be a successful practitioner in this field, one must develop skill in both the timing and applying pressure or force on the limb being tested. Muscle testing can be applied to any limb, but the most common is the arm extension or a leg flex test. The individual being tested is asked to extend an arm. The test can be conducted while the person is sitting, standing, or lying down. The examiner puts a hand on the extended arm or close to the wrist and asks the subject to resist when a downward pressure is placed on the arm.

Firm pressure is applied by the examiner using only a little force. Usually, the muscles of the subject will lock-in, and the subject easily resists the applied pressure. It is important for the examiner to slowly increase the pressure when muscle testing to allow the subject's nervous system to adapt and respond to the force placed on the arm. A rapid increase in pressure or too much force will overwhelm the resistance and push the arm down (See Figure 1). After the strong arm response has been established, any part of the body can be tested for imbalance or dysfunctions. The patient or examiner touches a specific part of the body while the arm muscle test is repeated. A strong response means no lesion or dysfunctions, but if the arm suddenly oscillates or drops, then there is a malfunction in the area being touched.

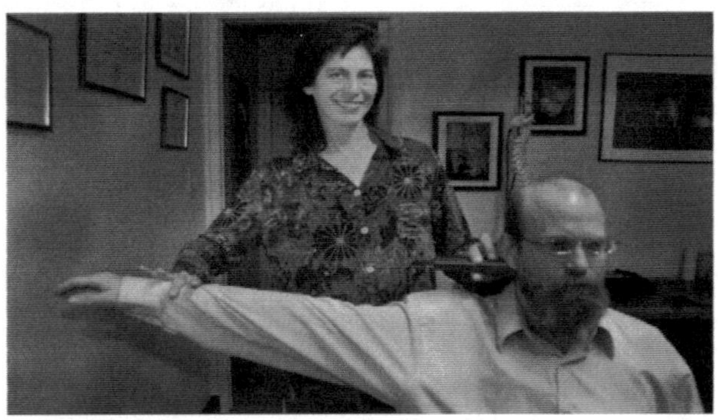

Figure 1. Determining neuromuscular dysfunction

Leaders in the field of natural medicine have found muscle testing to be valid, repeatable, and universal. In my experience, muscle testing has proved to be an accurate, non-subjective test for accessing the body's wisdom and for acquiring invaluable information about the state of the body. Because of the exquisite specificity of muscle testing, it allows one to identify individual vertebra that are not functioning optimally.

Muscle Test Using Extended Arm Method

In my research, I used muscle testing to determine whether a vertebra responds to vibrations – and indeed there was a response. I, along with several other chiropractors, conducted double-blinded and double-deaf studies using a series of 12 tuning forks in the range of 130.81 Hz to 246.94 Hz.

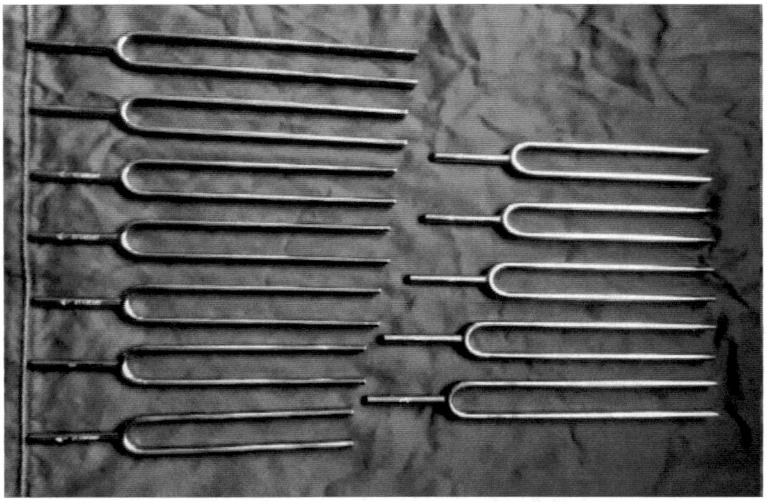

Figure 2. 12 different tuning forks used in testing muscle test response

The testing was arduous – each of the 24 vertebrae were tested with 12 different tuning forks (See figure 2). These tests revealed that each vertebra had a preferred frequency that generated a strong muscle test response. Non-preferred frequencies produced a weak muscle test response. My research revealed that the 12 semitones of the octave below middle C have a direct relation to the 12 bones in each spinal curve. I went on to do further research to determine whether the response was to a tone or a note. I used tuning forks with higher pitches such as C with a 256 Hz and 512 Hz. Unfortunately, at the time I did not have tuning forks in the range below 130.81 Hz.

A Set of 12 Stainless Steel Tuning Forks Ranging in Frequency from 130.81 Hz to 246.94 Hz

After years of research, I determined what seems to be the fundamental resonant frequency for each of the 24 "movable" vertebrae. What is even more important is that the research shows that applying the correct vibrational frequency to a vertebra can induce a sympathetic vibration that assists the restoration of the vertebra's "natural" frequency, or what I call the "home tone." When the vertebrae are restored to their optimal frequencies, harmonic waves set up a reverberation throughout the spine, creating a standing wave that maintains its functional and structural integrity.

Like a piano (and other instruments) the spine can go out of tune. Bones, like muscles, can lose their tone. Some factors, such as trauma, overuse, disease, and stress can diminish the tone of a bone. When the bones lose their tones the harmonic waves that travel through the spine become interrupted. This could lead to a spine that becomes inflexible with a diminished or exaggerated curve, and that affects not only the structure of the spine but the harmony of our nervous system. Bone is an excellent receiver and conductor of vibration and is capable of resonance.

Although I have used tuning forks as tools to transmit and restore vibrations, I soon learned that tuning forks were not a handy tool and often cumbersome to use repeatedly. Many tuning forks had to be applied to each vertebra and I found that impractical. Another inconvenience is that that the tuning forks had to be tapped or struck to get a vibration and repeated tapping was necessary for a sustained vibration. Not to mention the steel tuning forks I used (which are heavier than the aluminum forks) were awkward to hold and sometimes fell from my hand onto a subject's back. I imagined and dreamed of having a hand-held universal electronic tuning fork that would be able to quickly tune to precise frequencies and would be a much more efficient to use on the body. My first electronic tuning fork was based on the "sound bug," which was available commercially. The sound bug converts audio signals to vibrations. I hired an engineer to adapt the sound bug so it could be used as an electronic tuning fork.

The sound bug worked! But the vibrations were not as strong as I needed, so I continued my search for the right creative electronics engineer. After seven long years I found my Rudy. Overnight Rudy created what I dreamed of – an electronic tuning fork that became the *Song of the Spine Bone Toner.* It produces the frequencies of the spine and brain waves. After several more years of research, Rudy and I developed the Harmonic Resonator, an electronic tuning fork for the entire body (See figure 3). It generates the frequencies for the spine, spinal cord, brain, organs, fascia, chakras, and the solfeggio frequencies and Schumann resonance frequency.

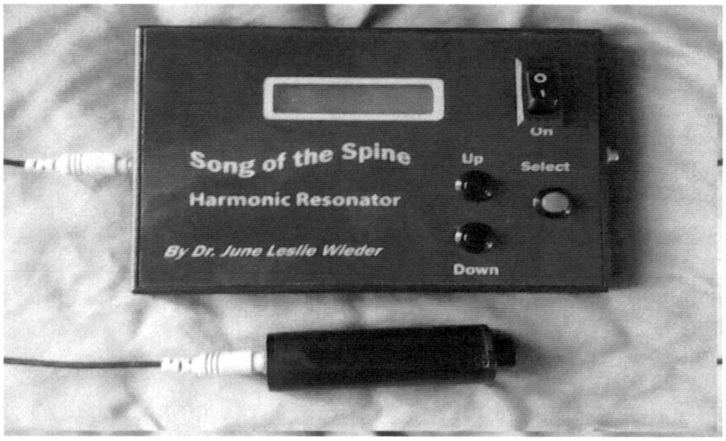

Figure 3. Harmonic Resonator Electronic Tuning Fork

A treatment usually consists of placing the vibrating probe of the electronic tuning fork on a specific vertebra, on the skull, or over an organ that is to be toned. The treatment last approximately 3 minutes for each spinal vertebra, brain wave, or organ that is toned.

Vibrating Probe of an Electronic Tuning Fork Is Placed on a Specific Vertebra for Toning

Recently I had a new machine built that is called the Adjustable Electronic Tuning Fork. This machine allows one to choose frequencies from 4 Hz to 2000 Hz. With this new tool I was able to re-test the frequencies for the *Song of the Spine* and found that the fundamental frequencies are approximately one octave lower than I originally determined. The fundamental frequencies range from 64 Hz to 122 Hz. which sounds like an open-stringed cello. The Adjustable Electronic Tuning Fork is suitable for both treatments and for research. Because it allows the selection of a specific frequency, the practitioner can test which frequency is most effective for a condition (see Figure 4).

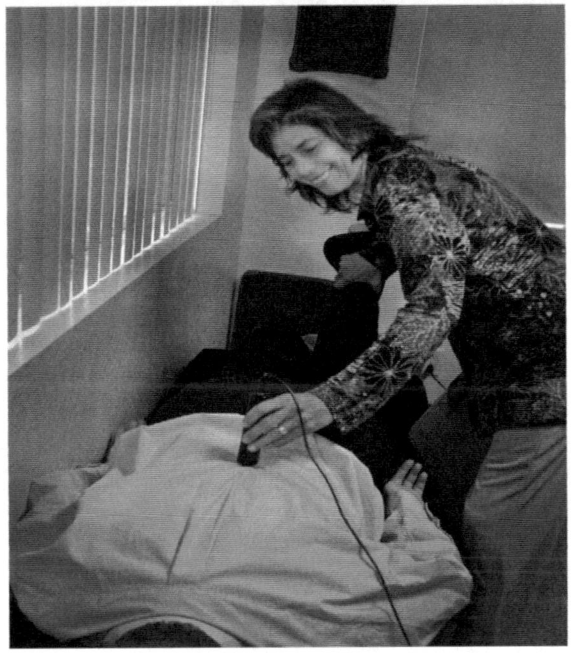

Figure 4. Testing for frequency in a patient's condition

I have found the electronic tuning fork to be very effective for many different conditions, such as musculoskeletal pains (both acute and chronic), sciatica, bulging discs, headaches, digestive disorders, breathing difficulties, frozen shoulder, autism, depression, anxiety, and post-traumatic stress disorder (PTSD). It is also effective for neurological conditions such as multiple sclerosis (MS). Future research may discover other conditions that can be successfully treated using the Adjustable Electronic Tuning Fork.

References

Wieder, J. L. (2022). *Song of the spine* website. Retrieved from http://www.songofthespine.com

Wieder, J. L. (2019). *The human symphony: Healing with sound and vibration* (N.P.).

Wieder, J. L. (2004). *Song of the spine.* North Charleston, SC: BookSurge Publishing.

Using Music to Alleviate Depression and Anxiety in Covid-19 Patients: Research Prospectus

David O. Akombo

Abstract

Objective: Existing evidence indicates that music can reduce the need for pain medication by reducing depression and anxiety, speeding the healing process, increasing endurance during exercise or physical therapy, and helping regulate pulse, respiration, and blood pressure in patients. The proposed research will be conducted to examine the effects of music on anxiety and depression on patients diagnosed with Covid-19 and have recovered from the disease. Coronavirus disease 2019 (COVID-19) is defined as illness caused by a novel coronavirus now called severe acute respiratory syndrome coronavirus 2 (SARS-CoV-2; formerly called 2019-nCoV) (CDC. 2019). Severe cases of COVID-19 may be associated with hypoxemic respiratory failure, acute respiratory distress syndrome (ARDS), septic shock, cardiac dysfunction, elevation in multiple inflammatory cytokines, thromboembolic disease, and/or exacerbation of underlying comorbidities. Covid-19 patients tend to experience anxiety and depression on varying levels. Although the effects of music on anxiety have recently been examined, very few studies have specifically examined the music effect on Covid-19 patients. Vulnerable and protected subjects including persons who have no English fluency or have a mental handicap that may prevent them from consenting and/or completing questionnaire will be excluded from the study. This study contributes to knowledge on the effects of music on pandemic-induced depression and anxiety. **Methods:** Patients (n = 500) will be randomly selected to participate in the study. Participants will include people aged 18 and above who have recovered from Covid-19 in a Latin American nation. Subjects will respond to a questionnaire examining their music listening habits during the period in which they were quarantined, isolated, treated and recovered. **Hypothesis:** (1) Music listening has an effect on depression in patients undergoing Covid-19 treatment during isolation. (2) Music listening has an effect on anxiety in patients undergoing Covid-19 treatment during isolation. (3) Music listening provides a measure of relief on patients undergoing Covid-19 treatment during isolation.

Keywords: Covid-19, quarantine, isolation, anxiety, depression, music

Introduction

Existing evidence indicated that music can reduce the need for pain medication by reducing depression and anxiety, speeding the healing process, increasing endurance during exercise or physical therapy, and helping regulate pulse, respiration, and blood pressure in patients. Corona Virus Disease 2019 (COVID-19) is an RNA virus, with a typical crown-like appearance under an electron microscope due to the presence of glycoprotein spikes on its envelope. Clinical features and risk factors are highly variable, making the clinical severity range from asymptomatic to fatal [Phan, 2020]. Severe cases of COVID-19 may be associated with hypoxemic respiratory failure, acute respiratory distress syndrome (ARDS), septic shock, cardiac dysfunction, elevation in multiple inflammatory cytokines, thromboembolic disease, and/or exacerbation of underlying comorbidities (COVID-19 Treatment Guidelines Panel, 2019). Music has also been shown to beneficially affect stress-related physiological conditions (Groen, 2007; Tan et al., 2008; Nilsson, 2008; Blood et al., 2001; Akombo, 2007).

Historical Background

Research on the effects of music and music therapy for medical patients has burgeoned over the past 20 years, examining a variety of outcome measures in a wide range of specialty areas (Blood, 2001; Akombo, 2007; Dileo et al., 2005; Beck, 1991). The literature suggests that increased anxiety leads to increased pain perception (McCracken, et al., 2009). During the Covid -19 pandemic, Ekberg (2021) has corroborated these findings where he notes:

> When we have the epidemic, they tell us to stay at home, and when we stay at home we don't have our normal social interactions... we worry about Covid and we don't get normal stimulation and joy in our life so we with worry we also get more anxiety and we get depression. (Ekberg, 2021, 21:55)

Intense experiences that change the music listener's state of consciousness and are accompanied by strong feelings of happiness have affected humans across the ages (Schäfer et al., 2013). Recent studies show that music can lower both state and trait anxieties thereby helping COVID-19 patients lower their pain perception (Akombo, 2007). Anxiety is a biologically mediated response to stress or change (May, 1977). Anxiety has been additionally defined as the apprehension cued by a threat to some value that the individual holds essential to his existence as a person. Its special characteristics are the feelings of uncertainty and helplessness in the face of danger (May, 1977; Selye, 1979). Anxiety is a common phenomenon occurring in hospitalized patients, a condition nurses deal with on a daily basis (Shuldham, Cunningham, Hiscock, & Luscombe, 1995).

Studies have revealed that listening to different kinds of music often leads to positive changes in individuals' lives (Schäfer et al., 2013). Positive and stress-free changes in individuals' lives are significant especially because prolonged experiences of stress and anxiety can lead to poor individual health as well as lower the quality of life. These can also trigger financial costs for families (McCaffery, et al., 1989). Therefore, developing cost effective anxiety and stress management approaches, like the implementation of music therapy, has become an important endeavor of current research efforts, and thus the purpose of the current study.

Subjects

Participants will be recruited from a small island located in the Caribbean. Participants (n = 500) will be conveniently selected to participate in the study through social media and mass media survey distribution. Participants will respond to a questionnaire examining their music listening habits during the period in which they tested positive for Covid-19 virus and the time they were quarantined, isolated, treated and recovered. Mass media texts will be sent by social media outreach (Facebook, Instagram, Twitter) which will include an enrollment packet consisting of a brief description of the study, an enrollment form on which the participants will be asked to indicate their willingness to participate, a consent form, and an initial demographic questionnaire. This initial demographic questionnaire will be helpful in the inclusion and exclusion processes. After they have responded to initial demographic questionnaire, those eligible will then receive the PHQ-9 and GAD-7 questionnaires. The PHQ-9 is a measure of depression while The GAD-7 is a standardized measure of anxiety. The two sets of the Patient Health Questionnaires (PHQ)-2 and PHQ-9 have been widely used for screening and diagnosis of depression in clinical and research settings. The GAD-7 (#5) is a standardized measure of anxiety.

Participants will be excluded if they have no English fluency, are mentally disabled, or pregnant. This study has been submitted for approval at The University of the West Indies Institutional Review Board in Barbados.

Measures Outcome Measures

I shall use a mixed-methods design by combining qualitative and quantitative techniques using questionnaires and ethnographies. The questionnaire design will entail two questionnaires, namely: The PHQ-9 (# 2) and GAD-7 (#5). The former is a measure of depression. The Patient Health is a diagnosis of guide for depression in clinical and research settings. The PHQ-9 was developed, in particular, to make criterion-based diagnosis of major depressive disorder (Arrieta, Aguerrebere, Raviola & Flores, et al., 2017). The GAD-7 (#5) is a standardized measure of anxiety. In a recent study, the instrument had performed better to identify generalized anxiety disorder (GAD) in comparison to other common measures (Plummer, Manea, Trépel & Mcmillan, 2015).

Fig. 1: Flow Chart of COVID-19 & Music Patient Population

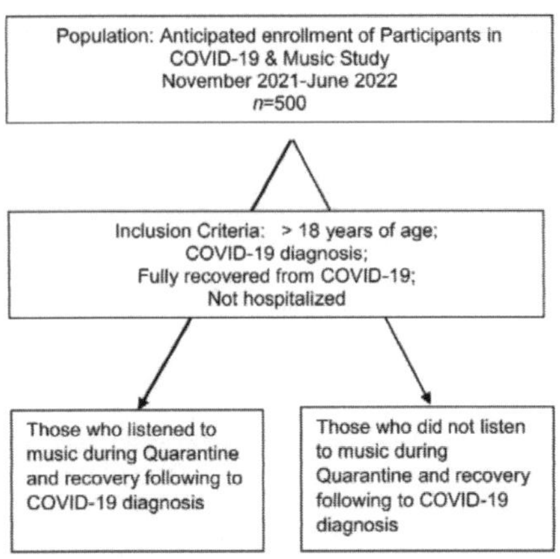

The participant's average daily depression will be measured using the PHQ-9 (# 2) and GAD-7 (#5). The participant's average daily pain will be measured using the Numeric Pain Rating Scale. The Numeric Pain Rating Scale (NPRS) is an 11 point (0-10) scale where higher scores indicate a higher level of pain (McCaffery, et al., 1989). I will ask the participants, "Using the numeric pain rating scale, what was your average level of pain each day?" The graphical scale depicts a ruler numbered from 0-10 indicating 0 as "no pain," 5 as "moderate pain," and 10 as "worst possible pain." This scale is widely used in patient care settings (Williamson, 2005; Ferreira-Valente et al., 2011). An illustration of the scale will be included in the Pain Diary for both the experimental and control group. Perceived pain for the experimental group will be self-assessed and recorded before the music listening session and immediately after the listening session using the NPRS. Pain medication usage will be measured using two questions where higher response scores reflect the presence of no medication usage for both questions.

Participants will be asked, “Did you take any pain medication immediately after your music listening session?” Responses for this question was, “Yes” (0) or “No” (1). With the second question, we will ask participants to describe the type of medication consumed, “Was it prescription medication or over-the-counter medication (eg. aspirin)?” Responses will include, “Prescription” (0), “Over-the-Counter” (1), or “I didn’t take any medication” (2).

Physician phone calls, physician visits, and Covid-19 ER visits with all patients reported outcomes will be recorded. Participants will be asked 3 questions regarding these behaviors, “Did you call your doctor’s office or clinic today because of breathing difficulties?” “Did you visit the Covid ER clinic today because of breathing difficulty?” and “Did you visit the ER (emergency room) because of pain?” Responses for these questions will be, “Yes” (0) or “No” (1).

Benefit in reducing pain will be measured using a single item. Participants will be asked “How helpful was the music listening in reducing your Covid-19 quarantine and treatment period?” The participants’ choices will include, “extremely helpful,” “very helpful,” “somewhat helpful,” “slightly helpful,” and “not helpful at all.” Benefit in Mood Improvement will be assessed using one question. Participants will be asked, “How helpful was the music listening session in helping you to relax, feel better, uplift your mood, etc.?” Responses will be recorded on a 5 point Likert scale that range from 1 = not helpful at all to 5 = extremely helpful for both questions.

Additional Measures

Participants will also be asked to record what genre or type of music they listened to and the length of their music listening session. Both questions will be open-ended. An additional question will be asked concerning participants use of headphones or earbuds during their music listening session.

Procedure

Over the span of 6 months, in 3 recruitment periods, 500 Covid-19 patients will be sent information regarding the study and the inclusion criteria for the clinical study. Participants who have fully recovered from (SARRS-CoV-2). Will be given a set of two questionnaires to complete. First, they will be given the Trait Anxiety Scale (T-Anxiety) which will evaluate relatively their general aspects of anxiety proneness. Then they will be given the modified (S-

Anxiety) which evaluate how they felt when they listened to music during their recovery from (SARRS-CoV-2). The Beck Depression Inventory (BDI) scale will then be administered to the participants to measure anxiety but with a focus on somatic symptoms of depression.

Statistical Analysis

Descriptive statistics will be used to determine the percentage differences in the anxiety and depression levels of those participants who listened to music during their illness to those participants that did not listen to music during this time period. Additionally, medication usage type, doctor calls, doctor visits, Covid-19 emergency room visits, and average daily depression and anxiety will be recorded and compared from this studies' participant and a control group. Descriptive statistics and t-tests will be generated using SPSS 23.0.

Results and Conclusion

The proposed research study will be conducted to examine the relationship between music listening and pain perception among a sample of COVID-19 patient. No study has been found that assessed the possible benefits of multiple days of music listening in patients who were experiencing COVID-19 symptoms. The COVID-19 pandemic brought about novel stressors requiring patients to apply a variety of coping strategies. In addition to reducing depression and anxiety, it is anticipated that the results will show a correlation between strategic music listening and a level of relief from depression and anxiety by patients diagnosed with COVID-19 disease compared to the levels experienced by those patients who did not listen to music.

References

Akombo, D., *Effects of listening to music as an intervention for pain and anxiety in bone marrow transplant patients.* Dissertation Abstracts International, 2007. 68(1-A).

Arrieta, J., Aguerrebere, M., Raviola, G., Flores, H., Elliott, P., Espinosa, A., Reyes, A., Ortiz-Panozo, E., Rodriguez-Gutierrez, E. G., Mukherjee, J., Palazuelos, D., & Franke, M. F. (2017). Validity and utility of the Patient Health Questionnaire (PHQ)-2 and PHQ-9 for screening and diagnosis of depression in rural Chiapas, Mexico: A cross-sectional study. *Journal of Clinical Psychology, 73*(9), 1076–1090. doi.org/10.1002/jclp.22390

CDC. 2019 Novel Coronavirus, Wuhan. China. CDC. Retrieved from https://www.cdc.gov/coronavirus/2019-ncov/about/index.html

COVID-19 Treatment Guidelines Panel. Coronavirus Disease 2019 (COVID-19) Treatment Guidelines. National Institutes of Health. Retrieved from https://www.covid19treatmentguidelines.nih.gov/.

Ekberg, S. (2021, October 8). What If You Stop Eating Sugar For 30 Days? [Video]. YouTube. https://www.youtube.com/watch?v=XS_kjTAnpP0

May, R. (1977). *The meaning of anxiety*. New York, NY: Norton.

McCaffery, M. and A. Beebe, *Pain: Clinical manual for nursing practice*. 1989, St. Louis, MO: Mosby-Year Book.

Phan, T. (2020). Novel coronavirus: From discovery to clinical diagnostics. *Journal of Molecular Epidemiology and Evolutionary Genetics in Infectious Diseases*, Volume 79. doi: 10.1016/j.meegid.2020.104211

Plummer, F., Manea, L., Trépel, D., & Mcmillan, D. (2015). Screening for Anxiety Disorders with the GAD-7 and GAD-2: A Systematic Review and Diagnostic Meta-Analysis. *General Hospital Psychiatry.* doi:10.1016/j.genhosppsych.2015.11.005

Schäfer, T. Smukalla, M. & Oelker, S. (2013). How music changes our lives. A qualitative study of the long-term effects of intense musical experiences. *Psychology of Music 42*(4), 525-544. doi:10.1177/0305735613482024

Selye H. (1956). *The Stress of Life.* New York: McGraw-Hill Book Company.

Selye H. (1979). *Stress of My Life: A Scientist's Memoirs.* New York: Van Nostrand Reinhold Company.

Shuldham, C., Cunningham G., Hiscock M., & Luscombe P. (1995). Assessment of anxiety in hospital patients. *Journal of Advanced Nursing, 22*, 87-93.

Contributors

Adina Dabija (LAc, MSTOM, Dipl OM) is a New York State Licensed Acupuncturist. She is nationally certified in Oriental Medicine by the National Commission for Certification of Acupuncture and Oriental Medicine, nationally certified by the American Board of Anti-Aging Health (ABAAHP), and a Certified Hypnotist. She holds a Clinical Master Degree from Pacific College of Oriental Medicine and formal certifications in Medical Qi Gong and Korean physical manipulations (Kyo Jung) from the Institute of Classical Asian Studies. With over ten years of clinical experience, Adina has been helping thousands of people from New York City to improve their health. Founder of several acupuncture clinics, Adina also created a holistic and meditation center in Port Washington, New York, with the focus on the top down (mind to body) approach in 2017. Since 2021 Adina is also working in holistic dentistry as a BioScan MSA Certified practitioner to prevent oral diseases. Adina has been teaching qi gong at Sands Point Preserve, Port Washington New York since 2018. Adina previously presented a pilot study with the title "Plato's Harmonious Man: a musical journey with-in using "charms", silence and remembrance of the cosmic harmony" at the ISQRMM 2019 Conference. Adina can be contacted at imilovan@unipu.hr

Antonios Chrysovergis (Tony Crisos) is a musician, educator, philosopher, and lecturer. He studied Music Performance at Berklee College of Music (BA) with special focus on jazz, and received his Masters in Music Education at Boston University. He has studied ancient Greek Philosophy, as well as Kabbalah, and now teaches and performs on both the guitar and the ancient Greek Lyre. While running a private home studio, he lectures throughout the country. He has also published several articles on music and philosophy. He can be contacted at tcrisos@me.com

Jonathan Seligman has been teaching choir, band, and general music for TK-6th grade at Casillas Elementary for 6 years, creating a new music community from the ground up. Academically, Seligman has received his B.Mus in Music Composition and a MA in Teaching from Point Loma Nazarene University, bearing a Single Subject Teaching Credential in Music and Special Education. He has furthered his learning, being fully Orff-Schulwerk certified as well as being trained in Feierabend's First Steps' and Conversational Solfege's methods. Accolades aside, Seligman's mission in all his groups is to teach beyond music literacy and discipline. While they are vital, his goal first and foremost is for students to be exposed to music of differing cultures and in tune with their musical expression. Jonathan can be contacted at jonathan.s.seligman@gmail.com

Rona Geffen is a Sound Researcher, Sound Healer, Music Producer, Composer, Vocalist, Sound Engineer, and Music Entrepreneur. The Alchemy of diffusing genres and practices was always present in Geffen's work as she studied music production and established a career as a Hardcore Pop-Avantgarde musician and multi-disciplinary artist. Geffen has released 3 solo albums, composed the electronic opera *STRIKE!*, and collaborated with artists such as Mad Professor, Ori Lichtick, Yasmeen Godder and many others. Having composed for dance and theater plays in Germany and Israel, Geffen also has participated in art and mixed media events worldwide and has toured with her solo act in Europe and USA. In 2015 Geffen started an independent sound research on the connection of geometry, mathematics, and sound with healing which led her to work with the high-end spatial sound technology 4DSOUND, the Spatial Sound Institute in Budapest, sound researchers, mathematicians, medical practitioners, scientists, and physicists. Geffen's research accumulated with her invention of Geometric Sound, a new property in the field of spatial sound, which showed promising indications in neurotrauma and stress rehabilitation. Today Geffen is the Lead Sound and Medicine researcher at The Works Research Institute focusing on the Geometric Sound effects on human health. Geffen is a certified sound medicine practitioner and studied various methodologies and instruments in sound and vibrational medicine including multi-instrumental & voice vibration medicine as integral medicine at the Sound Healing Academy (UK), vibrational acupuncture with planetary calibrated tuning forks according to Acutonics method (IL, UK), BioAcoustic technician at the Institute of BioAcoustic Biology and Sound Health (USA), and healing with singing bowls (IN/USA). She can be reached at booking@ronageffen.com

Sheerin Hosseini, PhD is a music educator from Boston, Massachusetts. She holds a PhD in music education from the University of Miami's Frost School of Music. Dr. Hosseini's research interests include music education and learning disabilities, Iranian music education, and pre-service teacher training. Sheerin can be contacted at sheerinmarie@hotmail.com

Dr. June Leslie Wieder has extensively studied the relationships between sound, movement, and form, and how they relate to health and healing. A classically trained pianist and now studying the cello. She has been a practitioner in the health field for more than 36 years. After earning a BSc in Psychology from Brooklyn College, she worked in the mental health field. She is a graduate of the Swedish Institute of Massage and has taught medical massage in California. In addition to her studies of various forms of bodywork, she was trained in the Trager Approach® personally by Milton Trager and has been a practitioner since 1986. She obtained her Doctor of Chiropractic degree from Life Chiropractic College West and started a private practice in New York City. She has been in private practice for over 26 years and currently has a clinic in Southern California. Her book, *Song of the Spine*, describes how sound can heal the spine (www.songofthespine.com) and her second book, *The Human Symphony*, expands vibration therapy to the entire body using the Harmonic Resonator, an electronic tuning fork. For more information go to www.thehumansymphony.comand www.drwieder.com. Dr. Wieder can be contacted at doctajune@gmail.com

David O. Akombo, PhD is Professor and Dean of the Faculty of Culture, Creative and Performing Arts at the University of the West Indies, in the West Indies. Intrigued by the peculiarities of music and its interdisciplinary confluence with medical and psychological practices, Dr. Akombo has studied and researched the effects of music on children and young adults both in schools and in hospitals. As scholar and performer, Dr. Akombo balances his research interests amicably as a music educator, ethnomusicologist, composer, and singer/drummer, having worked in Africa, The United States, and Southeast Asia where he studied with Balinese artists. His books,

Music and Healing Across Cultures (Ames, Iowa: Culicidae Press, 2006) and *Music and Medicine: Connections Found* (Seaburn, 2009) unfold the mechanics of the relationships between music, culture, physiological, and spiritual states in people from the anthropological and epistemological inquiry regarding music and healing. Dr. Akombo has worked jointly with the University of Florida's Center for Arts in Healthcare Research and Education (CAHRE) in which he collaborated in establishing the Arts in Medicine (AIM) program in Nairobi, Kenya. Located at Mater Hospital and funded by the Legislature and State of Florida Center for Cultural Affairs in Tallahassee, this is the first known Arts in Medicine program in Africa. Dr. Akombo continues to be active in music education research and interdisciplinary scholarship having conducted clinics in several countries. He is a 2017 recipient of the NSF Research Training Scholarship. Additionally, Dr. Akombo is member of the Music Educators National Conference, The National Social Science Association, and co-founder and President of ISQRMM. He can be contacted at dakombo@isqrmm.com

ISQRMM
C/O David O. Akombo, Ph.D.
Dean, Faculty of Culture, Creative and Performing Arts
P.O. Box 64 Bridgetown BB11000
Barbados, West Indies

ISBN 978-1-7138-6647-3

9 781713 866473

CURRAN ASSOCIATES INC.
proceedings.com